THE HUNT FOR RED OCTOBER

The smash bestseller that launched Clancy's career—the incredible search for a Soviet defector and the nuclear submarine he commands . . .

"BREATHLESSLY EXCITING." —*The Washington Post*

RED STORM RISING

The ultimate scenario for World War III—the final battle for global control . . .

"THE ULTIMATE WAR GAME . . . BRILLIANT."
—*Newsweek*

PATRIOT GAMES

CIA analyst Jack Ryan stops an assassination—and incurs the wrath of Irish terrorists. . . .

"A HIGH PITCH OF EXCITEMENT."
—*The Wall Street Journal*

THE CARDINAL OF THE KREMLIN

The superpowers race for the ultimate Star Wars missile defense system. . . .

"*CARDINAL* EXCITES, ILLUMINATES . . . A REAL PAGE-TURNER." —*Los Angeles Daily News*

continued . . .

CLEAR AND PRESENT DANGER

The killing of three U.S. officials in Colombia ignites the American government's explosive, and top-secret, response. . . .

"A CRACKLING GOOD YARN." —*The Washington Post*

THE SUM OF ALL FEARS

The disappearance of an Israeli nuclear weapon threatens the balance of power in the Middle East—and around the world. . . .

"CLANCY AT HIS BEST . . . NOT TO BE MISSED."
—*The Dallas Morning News*

WITHOUT REMORSE

The Clancy epic. His code name is Mr. Clark. And his work for the CIA is brilliant, cold-blooded, and efficient . . . but who is he really?

"HIGHLY ENTERTAINING." —*The Wall Street Journal*

UPDATED WITH NEW MATERIAL ON *SEAWOLF* AND *VIRGINIA*

"Takes readers deeper than they've ever gone inside a nuclear submarine."
—Kirkus Reviews

A rare glimpse inside a Los Angeles–class (SSN-688) nuclear submarine . . . with Tom Clancy as your guide.

Only the author of *The Hunt for Red October* could capture the reality of life aboard a nuclear submarine. Only a writer of Mr. Clancy's magnitude could obtain security clearance for information, diagrams, and photographs never before available to the public. Now, every civilian can enter this top-secret world and experience the drama and excitement of this stunning technological achievement . . . the weapons, the procedures, the people themselves . . . the startling facts behind the fiction that made Tom Clancy a #1 bestselling author.

SUBMARINE

INCLUDES:

- Exclusive photographs, illustrations, and diagrams
- Mock war scenarios and weapons-launch procedures
- An inside look at life on board, from captain to crew, from training exercises to operations
- The fascinating history and evolution of submarines

PLUS: Tom Clancy's controversial views on submariner tactics and training methods

continued . . .

The Bestselling Novels of TOM CLANCY

THE BEAR AND THE DRAGON

A clash of world powers. President Jack Ryan's trial by fire.

"HEART-STOPPING ACTION . . . CLANCY STILL REIGNS."

—*The Washington Post*

RAINBOW SIX

John Clark is used to doing the CIA's dirty work.
Now he's taking on the world. . . .

"ACTION-PACKED." —*The New York Times Book Review*

EXECUTIVE ORDERS

The most devastating terrorist act in history leaves
Jack Ryan as President of the United States. . . .

"UNDOUBTEDLY CLANCY'S BEST YET."

—*The Atlanta Journal-Constitution*

DEBT OF HONOR

It begins with the murder of an American woman in the
back streets of Tokyo. It ends in war. . . .

"A SHOCKER CLIMAX SO PLAUSIBLE YOU'LL WONDER WHY IT HASN'T YET HAPPENED."

—*Entertainment Weekly*

TOM CLANCY

SUBMARINE

A Guided Tour Inside a Nuclear Warship

Written with John Gresham

BERKLEY BOOKS, NEW YORK

SUBMARINE

A Berkley Book / published by arrangement with
Jack Ryan Enterprises, Ltd.

PRINTING HISTORY
Berkley trade paperback edition / November 1993
Berkley revised mass-market edition / January 2002

Visit our website at
www.penguinputnam.com

ISBN: 0-425-18300-9

BERKLEY®
Berkley Books are published by The Berkley Publishing Group,
a division of Penguin Putnam Inc.,
375 Hudson Street, New York, New York 10014.
BERKLEY and the "B" design
are trademarks belonging to Penguin Putnam Inc.

PRINTED IN THE UNITED STATES OF AMERICA

10 9 8 7 6 5 4 3 2 1

The author gratefully acknowledges permission for use of the following materials:

Photographs provided by the British Royal Navy, © British Crown copyright 1993/MOD, reproduced with the permission of the Controller of Her Britanic Majesty's Stationery Office; all rights reserved;

Photographs provided by the News Photo Division of the United States Navy; all rights reserved;

Photographs provided by John D. Gresham; all rights reserved;

Photographs provided by the Electric Boat Division of General Dynamics Corporation; all rights reserved;

Foreword courtesy of R. F. Bacon, copyright © 1993 by R. F. Bacon; all rights reserved.

This book is dedicated to
the families, friends, and loved ones of submariners,
who return that love, as well as
their love of God and country,
by going down into the sea in steel boats.

Acknowledgments

There is a popular quote that says "Failure is an orphan . . . but success has many fathers." If this book and the series that it starts turn out to be a success, it will be due to the vision and support of a great many people throughout the defense and publishing communities. First there is the team that helped me put it together. In the fall of 1987, I was introduced to a defense systems analyst named John D. Gresham. Over the years, we have had many lively discussions, and while we may not always agree, the disagreements always were thoughtful and insightful. Thus, I was pleased when John agreed to work with me as a researcher and consultant on this project. Backing up John and me was Martin H. Greenberg, the series editor. Marty's support in conceiving this book and the series, as well as his guidance of the entire project, have been vital. Laura Alpher, the series illustrator, created the wonderful drawings that reside in these pages. Thanks also go to Lieutenant Commander Christopher Carlson, USNR, Brian Hewitt, Cindi Woodrum, Diana Patin, and Rosalind

Greenberg for their tireless work in all the things that make this book what it is.

When we started this book, popular opinion around the Pentagon was that it could not be done. If any one person changed that, it was Vice Admiral Roger Bacon, USN (Ret.). As OP-02, he was instrumental in opening up the submarine community to the press and the public for the first time since nuclear subs started operating. Our special thanks go to him. In addition, Rear Admiral Thomas Ryan, USN (N-87), as well as Rear Admirals Fred Gustavson, USN, and Raymond Jones, USN, all provided high-level support. Lieutenants Jeff Durand and Nick Connally did yeoman work and tolerated dozens of ill-timed phone calls. In the Office of Navy Information, Lieutenants Don Thomas and Bob Ross just kept finding ways to make it happen. Special thanks to Russ Egnor, Pat Toombs, Chief Petty Officer Jay Davidson, and the staff of the Navy Still Photo Branch for all their tolerance and support.

Up at Groton, Connecticut, we want to thank Lieutenant Commander Ruth Noonan, USN, of the SUBGRU-2 public affairs office for her guidance during our visit. All around the base at Groton, the operators of the various trainers are to be thanked for allowing us to take part in a number of training exercises. Thanks also should go to the personnel and students of the submarine school. Also at Groton, we wish to thank Commander Larry Davis, USN, and the crew of the USS *Groton*, who opened their boat up to us, despite its being torn open for modifications and weapons loading. And to Commander Houston K. Jones, USN, and his crew on USS *Miami*, we pay the compliment of calling you "razors." From one side of the Atlantic to the other, those you have faced in exercises have only one thing to say: "Who were those guys?" Thanks also to the crews of USS *Greenling* and USS *Gato* for sharing their valuable training periods with us.

One of the great pleasures of doing this book was the opportunity to rekindle our friendship with the fine folks of Her Majesty's Navy. Rear Admiral Paul Fere, RN, and Commodore Roger Lane-Nott, RN, are to be thanked for their sponsorship of our project. Here in America, our way was paved by Rear Admiral Hoddinott, RN, Commander Nick Harris, RN, and Leading WRENs Tracey Barber and Sarah Clarke. At the Ministry of Defence, Commanders Ian Hewitt, RN, and Duncan Fergeson, RN, helped get us to the many places we visited. Mr. Ambrose Moore of the fleet public relations office in Northwood is to be thanked for his services as tour guide to bases in the U.K. We would also like to extend our thanks to the crew of HMS *Repulse*, who allowed us a brief visit into the world of the SSBN force. And finally, our warmest thanks go to Commander David Vaughan, RN, and the crew of HMS *Triumph* for their courtesy and friendship over several visits. Her Majesty can be proud of David and his men, for they have the same stout hearts as Drake, Nelson, and Vian.

Up in New York, our thanks to Robert Gottlieb and the staff at William Morris. And at Berkley Books, we owe a special debt to our editor, John Talbot. Thanks also to Roger Cooper for his patience and support of our work. Our personal thanks go to old friends Captains Doug LittleJohns, RN, and James Perowne, RN. Thanks also to Ron Thunman, Joe Metcalf, and Carlisle Trost for sharing their wisdom and experiences. And to Ned Beach, who taught us all to "run silent . . . and run deep." And lastly, our love to our families and friends, who tolerate our time away from them, so that we might tell our stories to the world.

Contents

Foreword

The transformation of Tom Clancy's wonderful fictional account of submarining in *The Hunt for Red October* to the reality of actual modern nuclear submarine capabilities and operations is long overdue. Now he brings a unique account of the nuclear-powered submarine, a vital component of naval power, to the public for the first time. This book explains the world of undersea warfare, from how people live within a steel tube for months at a time, to the many arrows a submarine puts in the quiver of national military power.

Twice in this century submarine warfare has threatened the existence of major powers. Submarines have always been a flexible and adaptable national asset, capable of many roles and missions. The submarines of World War I and II had some inherent stealth and could submerge to conduct attacks, but this property was limited by a lack of sustained power while under the sea's surface. The advent of nuclear propulsion made the submarine a truly stealthy platform. A so-called stealth aircraft can still be seen by the naked eye. A nuclear-powered submarine is truly invisible and not readily detectable. It is the original stealth machine and can remain undetected indefinitely. From this enduring covertness springs the awesome power of the modern submarine. Through the advances of ballistic and cruise missile technology the strategic nuclear deterrence mission and land attack capability have become an integral part of this military power. For decades the principal mission of a submarine has been to sink ships and submarines. Today, the nuclear-powered submarine's ability to affect events on land is one of its dominant features.

With Tom Clancy as our tour guide, let us view the submarine's history, its missions, the people and their families, the training, the boat itself with all its compartments and systems, and consider what these can do. If you spend years on the bridge of a submarine, as I have, you will notice how the dolphins that "ride" the crest of the exhilaratingly beautiful bow wave along the tear-shaped submarine hull do so at different positions for different classes or shapes of submarines. Why? I have always wondered. This tour you are about to take will come close to answering such questions, which are inherent to the mystique of a submarine.

I may not agree with all of the points present herein, but I do believe that upon completion of your tour you will understand why the submarine is the only naval platform that combines stealth, surprise, survivability, mobility, and endurance in a single unit. The employment of these characteristics provides a nation with a formidable maritime power, which should be understood by the public.

—Vice Admiral Roger Bacon, USN
Deputy Chief of Naval Operations for Undersea Warfare
January 1993

Introduction

Submarine. The very word implies stealth and deadliness. Of all the conventional weapons used by the world's armed forces these days, none is more effective or dangerous than the nuclear attack submarine (SSN). Since its creation in the United States some forty years ago, the SSN has become the most feared weapon in the oceans of the world. The modern SSN is a stealth platform with 70 percent of the world's surface under which to hide, its endurance determined not by fuel but by the amount of food that may be crammed into the hull, and its operational limitations determined more by the skill of the commander and crew than by external factors.

Understanding the capabilities of the modern nuclear-powered attack submarine requires a certain sophistication on the part of both a potential adversary and a visitor. Visually, a submarine is the least impressive of physical artifacts. Its hull does not bristle with weapons and sensors as do surface warships, and for one to see its imposing bulk, it must be in drydock. On those rare moments when a submarine is visible, this most lethal of ships appears no more threatening than a huge sea turtle. Yet despite that, the true capabilities of the modern SSN are most easily understood in terms of myth or the modern equivalent, a science fiction movie. Here is a creature that, like Ridley Scott's "Alien," appears when it wishes, destroys what it wishes, and disappears immediately to strike again when *it* wishes. Defense against such a threat requires constant vigilance, and even then, this will be ineffective much of the time. Thus the real impact of the nuclear submarine is as much psychological as physical.

In April 1982, the Monday after Argentina's seizure of the Falkland Islands, I happened to have lunch with a submarine officer and so got my first hint of what an SSN could do. The Royal Navy, my friend told me, would very soon declare that one of its boats was in the area of the disputed rocks. No one would be able to dispute the claim, which, my friend went on, would probably be false. "But the only way you know for sure that a sub is out there is when ships start disappearing, and that's an expensive way to find out." This is precisely what happened, of course. The mere possibility that the Royal Navy had one or more of its superbly commanded SSNs in the area immediately forced Argentina to reevaluate its position, and the Argentinean Navy, a lead player in the decision to seize the islands, was soon rendered impotent by its inability to confirm, deny, or deal with the mere possibility that an SSN *might* be lurking in the area.

As a practical matter, the Falkland Islands War was determined at that point. Ownership of any island is determined by control of the seas around it, and Argentina could not control the sea. The Royal Navy's SSNs prevented that, the first step in the RN's campaign to establish its own sea-control posture, making a successful invasion possible. The sinking of the cruiser *General Belgrano* was the unnecessary confirmation of what should have been obvious. While the nuclear-powered attack submarine may not be the most useful warship in the world since it cannot perform every traditional navy mission, it can deny an adversary the ability to execute *any* mission at sea.

"Here be monsters," the charts of ancient mariners used to say. They weren't right then, but current charts, especially those on surface warships, might profitably be marked to show that outside the thirty-fathom curve, yes, there be monsters. Nuclear-powered monsters.

The Silent Service

Early History

When tracing the roots of the modern submarine, one is usually faced with a number of different places to start. Legend has it that Alexander the Great descended into the ocean in 332 B.C. near the city of Tyre, in a primitive diving bell. The great mind of Leonardo da Vinci is said to have created a primitive submersible boat of wooden frame design covered in goatskins, with oars providing propulsion through waterproof sweeps. A British contribution to early submarine concepts came in the late 1500s from William Bourne, a carpenter and gunmaker. It included the concept of double hull construction, as well as ballast and trim systems. The first concept for a military submarine came from a Dutch physicist, Cornelius van Drebbel. In addition to actually building and demonstrating a primitive submersible, he proposed a design specifically created to destroy other ships.

It was the United States (albeit still colonies in rebellion) that created the first workable military submarine design. In 1776, a Yale University student named David Bushnell designed the appropriately named *Turtle*. The *Turtle* was an egg-shaped submersible boat that had the ability to sneak up on a ship, submerge under the intended victim, bore a drill bit with a waterproof time bomb attached into the bottom of the hull, and escape before the bomb was detonated by a clockwork fuse. It was propelled by a hand-cranked screw, and had room for one overworked crewman.

On the night of September 6, 1776, Sergeant Ezra Lee of the Continental Army took the *Turtle* to attack HMS *Eagle* of the British squadron blockading Boston. But when he maneuvered underneath, he was unable to attach his bomb. During his escape, he was followed by British soldiers in a rowboat. Frantic, he released the bomb, which exploded literally in the faces of his pursuers. Though all parties escaped unhurt, it was a promising start to the modern military submarine.

A more substantive advance was the *Nautilus*, designed by the American Robert Fulton, who would go on to design the first steamboat. The *Nautilus* was a distinct improvement over the *Turtle* in that it cruised under the intended victim, towing the explosive bomb or torpedo, as it was then called, until the bomb contacted the target and detonated with a contact fuse. The design was an exceptional success, destroying a number of target vessels in test runs. The French, who were sufficiently impressed to award Fulton a contract, actually considered for a time using it in the planned invasion of Britain. By 1804 Fulton was demonstrating the boat to the British, who despised the idea for its underhanded nature and, more importantly, its potential to sweep British ships from coastal zones. In the end, Fulton returned to America to begin work on his steamboats.

It remained for the Americans to create a submarine that would actually sink an enemy vessel in wartime. In 1863 a submersible boat was designed by Confederate army officer Horace Hunley. His boat, the CSS *H. L. Hunley*, was propelled by eight men turning a hand-cranked propeller. For armament, an explosive mine or torpedo was secured to a long spar protruding out in front of the *Hunley*. The idea was for the *Hunley* to ram the spar torpedo into the side of a target ship, where it would be detonated.

Unfortunately the *Hunley* was difficult to handle, and several crews, along with her designer, were killed during test dives. Nevertheless on October 17, 1864, the *Hunley* attacked the Union steam corvette *Housatonic* in the harbor at Charleston, South Carolina. In the ensuing attack the *Hunley* sank the *Housatonic*, although she herself was also sunk. A submarine had finally drawn blood in combat.

Over the next four decades a number of different submarine designs evolved in various European countries. In the 1880s a really practical design was built in America by an Irish immigrant, John Holland. Originally backed by the Fenian Society (an early North American free Ireland society), it was designed to allow Irish separatists to attack units of the British fleet. In 1900 Holland won a submarine design competition held by the U.S. Navy. From this contract came the USS *Holland* (SS-1), the first practical combat submarine. The *Holland* included such innovative features as self-propelled torpedoes fired from a reloadable tube, a battery-powered electric motor for submerged operations, and an advanced hull shape to allow it to move efficiently through the seas. The design was so successful that the U.S. Navy eventually bought a total of seven Holland-designed boats. Ironically, the British even bought some of the Holland boats for the Royal Navy. Holland's company, the Electric Boat Company, continues to build submarines as part of General Dynamics Corporation.

WORLD WAR I

The period before World War I saw a number of innovations in military submarines. This included the development of diesel engines, improved periscopes and torpedoes, and the development of wireless technology, which allowed them to be directed from shore bases. Within a month of the outbreak of World War I, the German *Unterseeboot* fleet, or U-boats as they came to be called, were sinking British naval units in the North Sea. In one well-known incident the elderly U-9 sank three British armored cruisers, causing over 1,400 casualties. Throughout the war, both the Allies and the Central Powers took a toll of each other's warships, especially in the Gallipoli Campaign in the Dardanelles.

During World War I the Germans consistently led the world in the production of new U-boats. But the international rules concerning attacks on merchant ships kept the Germans from fully utilizing their potential. Germany feared that unrestricted submarine warfare, with the practice of not warning the victim, might bring the United States into the war. By 1915 the need to isolate Britain from her sources of war supplies caused Kaiser Wilhelm to declare unrestricted submarine warfare an active policy. Soon German submarines were taking a huge toll of merchant shipping and threatening to win the war against Britain all on their own. But after the ocean liner *Lusitania* was sunk by U-20 in 1915, the United States entered the war on the side of the Allies. It would take two more years for the Allies to win the war and beat back the U-boat threat.

So important was the submarine in World War I that a whole new form of naval conflict, antisubmarine warfare (ASW), was born. From it came techniques such as the convoy and the Q-ship (armed merchant decoy), as well as weapons and sensors such as the antisubmarine detector (ASDIC/sonar), and the depth charge. And so deadly had the U-boats been that Germany was specifically banned from having them under the Treaty of Versailles. The victors of World War I split up the remaining U-boats for examination and testing. That might have been the end of military submarines except that the seeds of World War II were contained in the Treaty of Versailles, and the military submarine would continue to develop.

World War II

During the period between the world wars, submarine development continued at a steady pace. In the United States and Britain efforts were concentrated on the creation of long-range "fleet" submarines designed to support the battle fleets, while nations such as Japan, Russia, and Italy developed submarines more for coastal defense. Once Adolf Hitler had risen to power, Germany secretly began to rebuild its dreaded fleet of U-boats, in direct violation of the Treaty of Versailles. By the beginning of World War II, a number of improvements were made to the submarines themselves, such as torpedoes with magnetic fuses and sonars, and even small radar sets. And in Germany, the United States, and England, naval leaders had evolved very specific plans on how to best use these improvements.

By the outbreak of war in 1939, Germany had deployed her small fleet of U-boats at sea. Within hours, the U-30 sank the ocean liner *Athena*, signaling another round of unrestricted submarine warfare. Within a few weeks of the opening of hostilities, the U-boats had sunk a number of British warships and merchant vessels. The British responded with a series of patrols by their own fleet of submarines, damaging several German cruisers and sinking several U-boats. In addition, mindful of the damage inflicted upon merchant shipping in World War I, the British immediately instituted a system of transatlantic convoys and began to build up their ASW forces. But German fortunes soared with the capture of France and Norway in 1940, and once these prizes had been won, U-boats could be based much closer to the convoy lanes supplying Britain. The Battle of the Atlantic was on and would not be completely decided until the end of the war in 1945.

The Battle of the Atlantic was a battle of statistics: tonnage and numbers of ships sunk versus numbers of U-boats available and sunk. For Admiral Karl Dönitz, the German U-boat commander, it was a battle to get the greatest number of U-boats possible out onto the convoy routes. To do this he implemented what were called wolf pack tactics, setting a large number (ten to fifteen) of U-boats onto a convoy all at the same time. For a while, particularly during 1941 and 1942, the tactics worked. No less a figure than Sir Winston Churchill was reported to have said, "The only thing that truly worried me was the U-boat menace." He had much to be worried about, for Admiral Dönitz's U-boat force almost won the war by starving Great Britain into submission.

The British fought back though, using advanced tactics and equipment such as radar, escort corvettes, and frigates, and developing the small escort carrier.

In addition, the British had the ultimate secret weapon, Ultra. Ultra was the British program to penetrate German command communications, protected by the Enigma cipher system. Early in the war, with valuable contributions from the Poles and the French, England began to read an ever-growing flow of German messages. By 1941, through a combination of incredible technical analysis and outright theft of German cipher key books and captured Enigma equipment, the British were able to read virtually every message sent and received by the U-boats. Ultra allowed the British to route their convoys around known wolf packs and to start aggressively hunting the U-boats with aircraft and so-called hunter-killer groups. By 1943 the balance had turned decisively in favor of the Allies. Despite a number of German innovations such as the snorkel, homing torpedoes, and antisonar coatings, the battle was eventually won by the Allies.

In the Pacific, submarines actually won a major campaign against merchant shipping. In December 1941 Imperial Japan initiated a war of conquest against the Allies. At the start, things went very poorly for the United States. With most of their battleship force sunk or out of action after the bombing of Pearl Harbor, the only way the Americans could strike back was with their well-developed force of fleet submarines. It took a while to get rolling, especially when eighteen months were needed to repair a series of faults with the American Mark 14 torpedo and its magnetic fuse, but by late 1943 the American subs were beginning to make a real difference in the amount of material getting to Japan's war industries. Under the command of Admiral Charles Lockwood, the American boats were starting to starve Japan into submission. In addition, they were taking an increasing toll of Japanese warships.

By the end of the war in 1945, American fleet subs had sunk about a third of all the Japanese warships destroyed, and over half of the merchant ships. These successes did not come without cost. Over fifty U.S. boats had their epitaph written in the words "overdue and presumed lost." Along with the boats went some of the very best of the U.S. skippers, men like "Mush" Morton of USS *Wahoo*, "Sam" Dealey of USS *Harder*, and Howard C. Gilmore of USS *Growler*. Overall the U.S. submarine forces had the highest percentage of losses of any branch of the U.S. Navy. The American sub forces quietly paid in blood and boats for their victory, and earned for themselves a nickname that would stick: the silent service.

The Early Cold War Years

Almost as soon as the Allies won their victory over the Axis powers, another conflict, more sinister in character, started up between the Soviet Union and its former allies in the west. During the war the Russians had built the world's largest force of submarines. With the coming of what came to be known as the Cold War, they continued to build even further. For the next forty-five years the western allies, formed into NATO, lived in deathly fear that the USSR would flood its force of over three hundred submarines into the sea lanes. This threat—that the Russians could repeat or even better the performance of the Germans during the world wars—generated the main Cold War naval mission of the NATO forces, antisubmarine warfare.

The first decade of the effort was accomplished primarily by force of numbers. Despite the hopes that a decisive submarine technology would be found, none was. Improvements in submarine and ASW technology would evolve slowly. The major bottleneck was in the area of propulsion. Simply put, none of the different propulsion technologies—diesel, hydrogen peroxide, or gasoline—had ever provided the sustained high underwater speeds needed. The answer to this problem, though, was about to be found in the United States.

The Nuclear Revolution

The American propulsion breakthrough came from an unlikely source, a diminutive U.S. Navy captain named Hyman G. Rickover. Assigned after the war to the Navy's engineering branch, he was among the first to recognize the possibilities of creating small nuclear power plants that might be installed in submarines and surface ships. With these reactors, ships might steam tens of thousands of miles without refueling. For submarines in particular, it would mean freedom from having to come to the surface to obtain air for the diesel engines. In Rickover, and his newly created office of Director, Naval Reactors (DNR), the Navy had found the perfect blend of engineer, political insider, and bureaucrat to bring the first nuclear ships to fruition.

Submarines were Rickover's first priority, and a contract was let in the early 1950s for construction of the USS *Nautilus* (SSN-571) by the Electric Boat Division of General Dynamics. Utilizing a pressurized water reactor to produce steam for turbines, the design was successful beyond the wildest dreams of now-Admiral Rickover and the Navy. Considering that she was only a proof-of-concept vessel or prototype (the U.S. Navy has always considered its submarine prototypes fleet units, not research vessels), albeit armed with a full suite of weapons and sensors, the achievements of *Nautilus* and her crew were staggering. They dominated virtually every NATO exercise they participated in. In addition, in 1957 *Nautilus* became the first ship to transit the Arctic from the Pacific to the Atlantic, opening a whole new area for submarine operations.

Following the *Nautilus* came a second prototype, the USS *Seawolf* (SSN-575), powered by a liquid sodium reactor. Designed to achieve higher power output within a smaller volume, the reactor proved troublesome and was eventually replaced with one of the pressurized water type. In addition, the United States undertook production of a small class of nuclear boats (six) based on the design of the *Nautilus*. Named for the first unit of the class, the USS *Skate* (SSN-578), they provided a vast base of experience for operating nuclear submarines, as well as being extremely useful fleet units.

Skate herself made history by being the first submarine to surface at the geographic North Pole. Other prototypes such as the USS *Halibut* (SSN-587) and the USS *Triton* (SSN-586) explored the possibilities of using nuclear submarines to launch cruise missiles, and operating as a radar picket (to extend radar coverage for aircraft carrier groups). In 1960 *Triton* made history by becoming the first submarine to circumnavigate the globe submerged. Under the command of one of the U.S. Navy's best-known submariners, Commander Edward Beach (best known for writing the naval classic *Run Silent, Run Deep*), *Triton* duplicated the course of navigator Ferdinand Magellan some four centuries earlier.

The early U.S. nuclear boats were limited to a top speed of about 20 knots, submerged or surfaced.[1] These early boats had been built around conventional hull forms and were thus limited by the horsepower of their reactor plants

---

[1] To this day, the U.S. Navy will officially only admit that U.S. nuclear "submarines . . . operate at speeds over 20 knots, and depths over 400 feet . . ."

and the drag from their hulls. By this time the United States had experimented with a teardrop-shaped prototype diesel-electric submarine, the USS *Albacore*, which was able to reach submerged speeds of over 30 knots.[2] Combining the hull of the *Albacore* with Rickover's nuclear power plant, a new class of undersea hunter was born. USS *Skipjack* (SSN-585), the lead of a six-boat class, went to sea as the fastest submarine in the world. By 1960 the U.S. Navy had a fleet of nuclear submarines and a huge lead on the USSR and Great Britain, which had started their nuclear submarine programs later.

Along with the Skipjacks, another prototype boat was discreetly constructed to explore the possibility of a quiet SSN designed specifically to hunt other submarines. Named the USS *Tullibee* (SSN-597), she was the first SSN to have a large spherical sonar array in the bow, torpedo tubes amidships, and a quiet turboelectric drive system. And though she would have a history of engineering problems throughout her career (she was derisively known in Groton as Building 597), she introduced features that would be on every other class of SSN the United States has built.

Polaris Goes to Sea

Ever since the development of the first atomic weapons, the U.S. Navy had sought to develop a weapon system that would allow it to have a role in America's nuclear deter-

---

[2] Norman Friedman, *Submarine Design and Development*, U.S. Naval Institute, 1984.

rence mission. Initially the Navy used carrier aircraft that could deliver the early nuclear weapons on one-way missions to their targets. What the Navy really wanted was to merge the new technologies of ballistic missiles, smaller thermonuclear weapons, inertial guidance systems, and nuclear submarines into a single weapon system. The program was called Polaris, and it became the top U.S. naval weapons development program of the 1950s. Pushed aggressively by Admiral Arleigh Burke, the U.S. Chief of Naval Operations, and managed by an authentic programmatic genius in Rear Admiral "Red" Rayborne, the program moved forward at an amazing pace. By the late 1950s a small, reliable missile known as the Polaris A1 was ready to have a platform built for it. The problem was that submarine construction takes time, and the United States wanted to deploy the Polaris by 1960.

To accomplish this, Admiral Rickover had Electric Boat split one of the Skipjacks under construction (she was the original USS *Scorpion*) just aft of the sail and insert a plug containing sixteen Polaris launch tubes as well as all the missile launch controls and maintenance equipment. Christened the USS *George Washington* (SSBN-598), she would be the first of a five-boat class of fleet ballistic missile (FBM) submarines that would become the most powerful deterrence force in history. When the *George Washington* successfully test-fired two of the Polaris A1 missiles on July 20, 1960, off Cape Canaveral, Florida, the system became operational. Later that year she left on the first of what has become over three thousand FBM deterrence patrols, each lasting roughly sixty to seventy days. After each patrol, the onboard crew switches with a second crew, alternately known as "blue" and "gold," so that the high operational tempoes (time on patrol) can be maintained. So successful has the fleet ballistic missile program been that it is reported no U.S. FBM boat has ever been tracked for any duration. Thus the silent service entered a new era and added to their already formidable reputation. Within a year, a second batch of five missile boats, led by the USS *Ethan Allen* (SSBN-608), was on order.

The Quiet Revolution

Following the Skipjack and George Washington–class boats, the United States embarked upon a new direction in nuclear submarine development. It was decided, after an analysis of early Soviet nuclear boat characteristics, that high speed (over 30 knots) was not necessarily desirable. Submarines traveling at high speed make a great deal of noise, which can be heard by other submarines and surface vessels. Thus diving depth and quietness rather than speed would become the qualities that characterized the American submarine designs of the 1960s.

The first of the new deep-diving/quiet boats was to be the USS *Thresher* (SSN-593). Unfortunately, during rectification trials off Nantucket in 1963, the *Thresher* was lost with her entire crew as well as several civilian and U.S. Navy "riders." In the investigation that followed, it was determined that a brazed piping joint in the engineering spaces may have weakened during the shock trials and burst, causing massive flooding that prevented the boat from surfacing. The Subsafe program was later instituted by the U.S. Navy, which developed the deep-submergence rescue vehicle (DSRV) to rescue the crew of a sunken submarine. The class was continued, named after the next boat in line, USS *Permit* (SSN-594).

The Force Expands

As the 1960s drew on, the U.S. Navy began a vast expansion of its nuclear submarine program. The plan was to build an additional thirty-one SSBNs as well as a new class of attack submarines. The ballistic missile boats would be armed with a new generation of ballistic missile, the Polaris A3, with a 2,500-mile range. In addition the SSNs were to be armed with the SUBROC, a submarine-launched rocket with a fifty-mile range and a nuclear depth charge capable of destroying enemy submarines. All this was part of the military buildup originally proposed by President John F. Kennedy and carried out by the administration of President Lyndon B. Johnson. First on the list were the new FBM boats, or "boomers," as they were being called.

Starting with the basic plan of the USS *George Washington*, the designers sought to install all the quieting technology that had been incorporated into the Permit-class boats. In addition they made the missile section large enough to accommodate not only the new Polaris A3 missile but a new missile that would have superior range and multiple warheads, the Poseidon C3. Named for the lead boat in the class, USS *Lafayette* (SSBN-616), these boats were most impressive for their numbers built—thirty-one in all—and their stealth. And with the ability to upgrade their missile battery to the Poseidon C3 when it came on line in the 1970s and the Trident C4 in the 1980s, these boats were going to have a long service life. (As this book goes to press, about a third of the Lafayette-class boats are still in service.)

After the Lafayette program was underway, the Navy turned its attentions to the problem of an improved attack boat. Again, analysis of the submarines being produced by the USSR showed that deep-diving quiet boats were best. The lead boat of the new class was USS *Sturgeon* (SSN-637). Much like the Lafayette-class nuclear ballistic missile submarines (SSBNs), this class was characterized by a relatively large production run—thirty-seven units—and reduced noise signature. This improvement did not come without cost though, as the top speed of the Sturgeon-class boats was down to around 25 knots.[3] Nevertheless they proved to be superb boats with excellent capabilities and were, along with the Permit-class and Skipjack-class boats, the backbone of the U.S. attack submarine force.

In the midst of all this growth and success in the submarine force came a tragedy. In 1968 one of the Skipjack-class boats, the USS *Scorpion* (SSN-589), went missing while returning from a regular patrol in the Mediterranean. For the first time in modern U.S. submarine operations the words "overdue and presumed lost" were used to inform the world of a possible SSN loss during normal patrol operations.

While the exact method of location is still not openly known, it appears that the U.S. seabed-based sound listening (SOSUS) network heard an explosion from *Scorpion*. Later that year a survey expedition, utilizing the bathyscaphe *Trieste*, located the wreck near the Azores, relatively intact on the seabed. It was concluded her loss may have been due to an internal explosion, though the exact cause has never officially been announced.[4]

On a more positive note, the Navy built several new prototype submarines to explore new propulsion technologies.

---

[3] Patrick Taylor, *Running Critical*, Harper and Row, 1986, p. 58.
[4] Ibid., p. 259.

The USS *Glenard P. Lipscomb* (SSN-685) was designed to look again into the feasibility of using a turbine-electric drive, while the *Narwhal* (SSN-671) carried a prototype reactor using natural circulation rather than pumps, which can be very noisy, to move coolant through the reactor system. While they did provide useful data for future submarine designs, neither boat was considered to be particularly successful. With this lack of a propulsion breakthrough, the stage was set for the fight over the design of the next generation of nuclear submarines.

The New Generation of Boats

In the late 1960s, the U.S. intelligence community began to receive disturbing indications that the nuclear submarines of the Soviet Union had much higher performance capabilities than previously thought. A debate broke out between Admiral Rickover at the Naval Reactors Branch and the Naval Sea Systems Command (Navsea) over the direction of the next generation of attack submarines. Rickover felt that what was needed was a quiet, high-speed (over 35 knots) attack submarine able to support the carrier battle groups deployed by the U.S. Navy. Navsea was supportive of a design called Conform, utilizing a natural circulation reactor, which would recover the speed loss of the Permits and Sturgeons (down from 30 knots to 25 knots) and improve the radiated noise levels.[5] Eventually Rickover won out, and a twelve-ship class, its lead boat to be named USS *Los Angeles* (SSN-688), was planned, with Electric Boat as the prime contractor.

---

[5] Ibid., p. 58.

The Los Angeles–class boats delivered their promise of high speed as well as being the quietest attack submarines ever created up to that time. The price they paid for that speed was that their hulls were thinned; they could dive only to about three-fourths the depth of the Sturgeon and Permit classes (approximately 950 feet/300 meters).[6] In addition habitability suffered, with a greater percentage of the crew having to rotate bunks (called "hot bunking"). Finally, the Navy and Electric Boat had significant financial and program management problems, along with a desire to expand the class more quickly, leading to a second-source contract for construction to Newport News–Tenneco. In spite of this, the first Los Angeles–class boats came on line in the late 1970s and immediately set new standards for quiet operations and speed. Some sixty-two Los Angeles–class boats would eventually be contracted, making it easily the largest class of nuclear submarines ever built.

In addition a whole new series of submarine weapons came on line in the late 1970s and 1980s, including the new Mod 4 and ADCAP versions of the Mark (Mk) 48 torpedo; the UGM-84 Harpoon antiship missile; and three separate versions of the R/B/UGM-109 Tomahawk missile for nuclear land attack, antiship use, and conventional land attack. All of these new weapons, combined with the addition of a vertical launch system and stowage for twelve Tomahawk missiles on the Los Angeles–class boats, suddenly made U.S. SSNs capable of a whole range of missions that Admiral Rickover had not dreamed of when he first pushed through the proposal for *Nautilus* in the 1950s.

The new class of boomer was somewhat clearer to de-

---

[6] Ibid.

sign: the primary criterion was stealth. When the first boat of the new class, the USS *Ohio* (SSBN-724), appeared, she was reported to radiate less noise than the surrounding ocean and surface traffic, making the Ohios the quietest submarines ever to take to sea. Another major improvement was the number of missiles carried. All previous SSBNs produced by the United States had sixteen missile tubes. The Ohio class has twenty-four missile tubes, with a diameter large enough to accommodate not only the Trident C4 missile (the replacement for the Poseidon C3), but also the Trident D5 missile. The Trident D5 had significant improvements in both range and accuracy, making it the most powerful component in the U.S. nuclear arsenal. Under the terms of the START-II treaty signed in 1991, the bulk of the U.S. strategic nuclear strike power will be carried on the Ohios.

The Next Generation

With the coming of a new series of arms limitation treaties (the START series), the United States does not have any plans to build a new class of SSBNs. In fact, the Ohios were built with enough growth potential in their design that service lives of thirty-five to forty years are entirely possible, and if replacements are required, they won't be needed until around the year 2015.

Attack boats are another thing entirely. A follow-on to the Los Angeles class has been planned for some time, and the lead boat of the new class, USS *Seawolf* (SSN-21), is due to come on line in the late 1990s. The Seawolf design makes good virtually all the shortcomings of the Los Angeles–class boats, particularly in the areas of depth (back to approximately 1,300 feet/400 meters), habitability (improved crew comfort), and weapons load (a combination of fifty weapons).[7] Such things come at a severe cost though, both in money and size. Seawolf is huge, over 9,100 tons displacement, making it the largest attack submarine in the world other than the Russian Oscar-class guided missile boats. And with a cost at this writing of over $2 billion per copy, the Seawolf production run is currently limited to only two units.

As production of the Los Angeles and Ohio classes winds down, and with the Seawolf program being terminated early, the future of the U.S. nuclear submarine force is in doubt for the first time in forty-five years. What has been the premier weapons system of the Cold War now seems to be a system in search of a mission and an audience. We will explore the future later on, but first let's look at the present, and what the taxpayers have bought for themselves and their silent warriors.

---

[7] A. D. Baker, *Combat Fleets of the World*, U.S. Naval Institute, 1993, pp. 809–811.

Building the Boats

It sounds so simple: building the boat. Yet this is a process that starts years before the submarine enters the fleet. Remember, in 1969 the U.S. Navy was considering the design of the Los Angeles–class submarines, which began to enter the fleet some seven years later. Even today, if you could order one (the line is being shut down to produce the Seawolf-class boats), it takes six years from contract signing by the Naval Sea Systems Command (Navsea) in Arlington, Virginia, until the completed boat is commissioned into the force. This process starts in the steel mills of the eastern United States and the computers of the Electric Boat Division of General Dynamics. It also starts in the cities and towns of America, where the raw materials for the crews are born, raised, and educated. Let us take a quick look at how it is all done.

The Sharp Edge—The Crew

It's hard to separate the steel and electronics of the boat from the flesh and blood of the men who will serve as her crew. In a manner of speaking, the crew is a part of that machine headed to sea. I suppose if robots could do the job of men, they would have taken over the submarine force by now. But the day when a robot can survive the shock of an explosion, the rush of flooding water, and have the cunning of a man is still years away. And until that day comes, men will go into the sea in the steel cylinders called submarines.

Where the crew come from is, quite simply, everywhere. From every town and village, from the largest inner city, the suburbs, and the rural countryside. What motivates each of them is probably a little different. For Admiral Chester Nimitz, the World War II Commander in Chief of the Pacific and himself an early submariner, it was the desire to see a body of water larger than the mud puddles of west Texas. For some who want submarines, it is the desire to work on one of the most powerful and sophisticated pieces of machinery ever built. Others see the Navy and the submarine service as a way out of the poverty and despair of whatever situation they may have been born into. Whatever the reasons, they have all come to the Navy to find something to build their lives around.

Let's say that a young man who has graduated from high school wishes to join the Navy and "see the world" from the voyages of a submarine. That young man (sorry, ladies—men only on subs at the time this book is being written) would probably apply at his local recruiting office. From here he is transported to the local personnel recruiting depot for basic training. Some weeks later, he moves on to his specialty—electronics, sonar, machinery, etc.—or

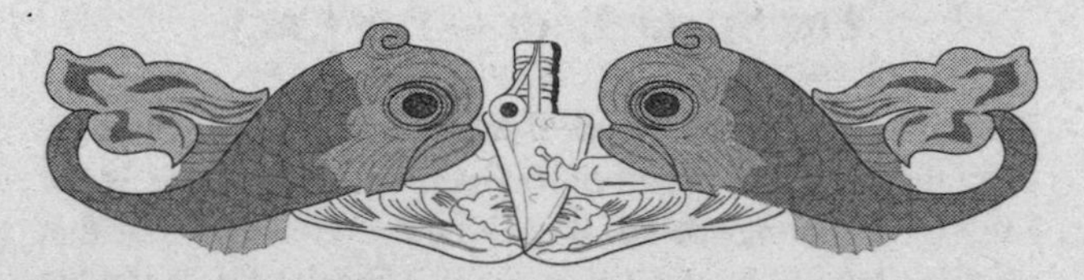

U.S. "Dolphin" submarine logo. *Jack Ryan Enterprises, Ltd.*

"A" school, which gives him the skills necessary for his job when he joins the boat. If he has decided to select nuclear power as his specialty, he goes to six months of nuclear power school (NPS) in Orlando, Florida, followed by six months of training on one of the nuclear reactor prototypes. Assuming that he has selected submarines as his service, the young recruit is next headed to the home of the submarine, the U.S. Navy Submarine Base in Groton, Connecticut, to attend Submarine School. Sub school teaches the recruit the basics of what he needs to know about life aboard submarines. From here he moves onto the crew of a boat for his first tour, which will probably last a couple of years.

One of the advantages the submarine service has in attracting the cream of the Navy's new recruits is money. Ordinarily a new sailor who selects nuclear power as his specialty would be given the rank of seaman apprentice, but the submarine service immediately makes a new recruit a petty officer. This is important because of the pay differential. While it might not look like much of a difference, it can be enough to let a young man get married so that he can start and support a family. The submarine service asks much of the young men who drive their boats, and the need for every sailor to have a home and someone in it is a cornerstone of their tradition.

Once onboard his first boat, the new crew member's first major career task will be to qualify for his "dolphins," which certifies him as a submariner. From there, he is expected to take his qualification boards and move up the promotion ladder. After this first tour, if he chooses to reenlist (and many do) he will probably be given the opportunity to move to one of the various schools as an instructor. This might be at one of the reactor prototypes or the firefighting school in New London. Wherever it is, he will be asked to put back into the new recruits some of the knowledge and experience he has gained. And this is the cycle that he will follow for most of his career.

Qualify and earn promotion, that is the key. Eventually the submariner might be given the chance to become a warrant officer, or perhaps go to college to become an officer, or "mustang," as they are known in the Navy. For those choosing to remain as enlisted men, the ultimate honor is to make the rank of master chief, who is usually given the title Chief of the Boat, or COB, on a submarine. This position is considered the equivalent of the executive officer (XO), in charge of the enlisted men on a boat. These are frequently well-educated men with graduate degrees. And to say that the commanding officers (COs) of submarines respect their opinions is something of an understatement. If anything set our service apart from that of the former Soviet Union during the Cold War, it was the cohesion or "glue" that our noncommissioned officers provided the Navy. They are the keepers of what corporate America might call corporate memory or tribal knowledge, or what in the Navy they just call tradition.

The route of an officer is somewhat different from that of the enlisted men. For starters, the Navy is rather particular about who gets to drive their nuclear boats. So while the

Navy might be satisfied with a psychology or history major driving an F-14 Tomcat or Aegis cruiser around the block, for their nuclear officers they want engineers. Or, more correctly, university graduates with hard science degrees. There are several ways for a young man to get into this career path. Certainly the most conventional route is the U.S. Naval Academy at Annapolis, Maryland. There also is the Reserve Officers Training Program (ROTC) in place at many U.S. college campuses. This four-year program helps provide tuition, books, and a small monthly stipend to help support the young man, who is commissioned an ensign when he graduates. The final way for college graduates is just to volunteer through the Officers Candidate School (OCS) program. In this case they will be put through a three-month training program, hence their nickname of "ninety-day wonders," after which they are also commissioned as ensigns.

The first step on the road to becoming a U.S. Navy submarine officer starts with selection by the Director, Naval Reactors (DNR-Navsea Code-082E). This involves a series of personal interviews with the DNR (a four-star admiral no less) to assess the candidate's technical knowledge and ability to handle stress. When Admiral Rickover used to handle these interviews, the questions took on a sometimes bizarre and personal nature, but as people in the submarine community will tell you, it seems to have produced a *very* capable corps of submarine officers. At this point the new submarine officer heads off to a year at NPS and the reactor prototype schools.

Once this is completed, he will be sent to the Submarine Officers Basic Course (SOBC) at Groton, Connecticut. SOBC takes three months and is roughly equivalent to the enlisted men's Submarine School course. Upon completion of the SOBC, he finally is assigned to his first boat, where he will probably spend the next two to three years. Much like his enlisted counterparts, he will spend much of his time standing watches and qualifying for his "dolphins." He will also be assessed in his ability to handle and lead the men assigned to his division and watches. Even at this stage of a young officer's career, he is being tested for his ability to command a boat in the future. During his first sub tour he will take the engineer's exam, again supervised by personnel from DNR. This is a critical exam because it is the first major stay/leave criterion, allowing him to stay in submarines or pointing him to some other part of the Navy. Success means that the officer is now qualified to be assigned as chief engineer of a boat. From here he will probably do a shore tour on staff at a sub squadron or as an instructor at one of the schools. He also will probably have been promoted to lieutenant by now.

After the shore tour the officer, now not so young, returns to the submarine school at Groton for another six-month training course. This one, known as the Submarine Officers Advanced Course (SOAC), is designed to prepare and qualify the officer as a department head—engineering, navigation/operations, weapons, etc.—on a boat. It is also one of the required steps on the road to command of a boat. Now the officer heads back to a boat for his three-year department head tour. By now a senior lieutenant, he is ready to screen for the *big* step on the road to command of his own boat, becoming an Executive Officer (XO). After he has screened for XO, his next training course is the three-month Prospective Executive Officers (PXOs) course, which qualifies the officer for his tour as Executive Officer of an SSN or SSBN. If he successfully completes his XO tour, he will probably head for a shore tour, possibly in one of the many joint billets that are considered so important to the career of American military officers. From here he is selected for the rank of commander, screens for command, and heads to the Prospective Commanding Officers (PCO) course and, finally, to command of his own boat.

This last step, the PCO course, should not be thought of lightly. Much has been made of the U.S. Navy's fixation with nuclear reactor safety when selecting skippers. A good record with power plants is certainly one of the major selection criteria for command. The Navy feels, probably with good reason, that they must have a *perfect* operating record for the American public to allow them to continue operating ships and submarines with nuclear power. With this said, though, it is the PCO course that actually qualifies a man to command one of the U.S. Navy's boats and not the scores on his engineering exams.

The PCO course was created in 1946 by James Forrestal, then Secretary of the Navy and later Secretary of Defense. It allows the submarine service to have total autonomy in the selection and training of its submarine skippers. Certainly, advanced training programs like Top Gun—for U.S. Navy and Marine fighter pilots—Red Flag—for U.S. Air Force aircrews—and the National Training Center—for U.S. Army units—are better known to the public, but the submarine PCO course is easily the equal of any of these. Successful completion of the PCO course is mandatory if a man is ever to command a U.S. nuclear submarine. Another aspect of the PCO course that is not generally known is exactly what the curriculum consists of. For the record, each course, which is approximately six months long and enrolls between ten and twelve officers, teaches them the tactical and operational intricacies of commanding a U.S. nuclear submarine.

During the ensuing six months, the prospective CO will practice approaches and fire something like five to seven "live" (exercise) weapons (Mk 48s, Harpoon and Tomahawk missiles) under a variety of conditions. The course curriculum is both wide and varied, with improvements and changes being made after each and every course. The challenge for the instructors of the PCO course is that in just a dozen years, they have gone from a course with only one primary weapon (torpedoes) and mission (ASW), to having the broadest range of missions—ASW, antishipping, mining, strike warfare, intelligence gathering, etc.—and weapons—torpedoes, missiles, and mines—in the entire U.S. Navy. And as in the submarine qualification courses of other countries, especially the Royal Navy's Perisher course, any miscue or mistake can be reason enough for an officer to be disqualified.

At the end of the six months, if he has completed all aspects of the course, and if the instructor feels he is both qualified and ready, the PCO student graduates. At this moment he will have achieved the goal of every submarine officer, command of his own boat.

BOAT CONSTRUCTION

Let me try to give you the condensed version of how an Improved Los Angeles (688I) is built.

The first step in the process is for the Navy to decide that they want to build a boat. This decision is made in the Undersea Warfare Office of the Office of the Chief of Naval Operations (OPNAV). Until recently this office was known as OP-02 and was headed by Vice Admiral Roger F. Bacon, USN. In November 1992, through an OPNAV reorganization, this office was renamed N-87 and is now headed by Rear Admiral Thomas D. Ryan, USN (Director, Undersea Warfare Division). It is here that the requirements for such boats are established and the request for proposal is developed. This is usually done in batches or "flights" of boats to a particular shipyard. For our purposes, we will assume that the builder is the Electric Boat Division of General Dynamics Corporation. Their yard at Groton, Connecticut, would submit a bid to Code 92 (attack submarines) at Navsea, and after a series of negotiations, the contract to build the boat would be awarded. From here the funding for the boat would have to be submitted in the president's defense budget, approved by Congress, and the money allocated in the federal budget.

Once the boat has been approved, the actual process of construction begins. The first step in the process is to order items with long lead times, like the nuclear reactor, and heavy machinery, like reduction gears and turbines. The reactor, in this case a General Electric S6G, is ordered and supplied as a piece of government-furnished equipment by Code-082E at Navsea, the Office of the Director of Naval Reactors (DNR).

A year or two later, when these items begin to show up at Electric Boat—known simply in the Navy as "EB"—the actual construction of the boat begins. The first step is the construction of the pressure hull. EB manufactures its own pressure hull barrel sections in a special facility at Quonset Point, Rhode Island, which takes three-inch-thick hardened steel plates and works them into the curved sections. The sections are carefully welded together to make up the barrel sections, which are barged to the EB yard at Groton. The work now proceeds to the huge building shed at EB. Here the hull sections are welded together into a single long cylinder to form the pressure hull. It is miserable work, with the metal of the barrel sections having to be heated to 140°F/64°C just to prepare for welding. Each section is then hand-welded to the next by men often on the verge of heat prostration, exhaustion, and dehydration. Men must do this work because no machine can do the job to the standards of Navsea and DNR, and even this work must be checked by Navy inspectors armed with mirrors and X-ray machines. The individual sections of the hull are packed with items that are too big to install later, such as the reactor, torpedo and vertical launch system (VLS) tubes, and the turbines.

Once the cylinder of the pressure hull is finished, it is moved down the production way to have the machinery mounts, trim tanks, and internal deck structure installed. Now more and more components of the boat are delivered to the yard. Also during this time the first elements of the precommissioning unit (ships and submarines are known as "PCUs" before they are commissioned as "USS") crew begin to arrive at EB. These are the Navy personnel who will first take the new boat to sea. Usually the initial cadre is composed of a few officers, including the commissioning CO, and a number of chiefs. Their job will be to oversee the final fitting out of the boat, as well as being the Navy's representatives to EB for the commissioning. Eventually the ends of the hull are sealed with end caps, and the superstructure is installed.

When the last of the heavy structures like the conning tower/fairwater are installed, and the hull is declared water-tight, it is time to roll the boat out of the building shed and launch it. By this time, the PCU crew has been completely assigned, working day to day with the EB personnel. Once the boat is launched, it is towed to a dock where the rest of the sub's equipment will be installed and tested. This can take between six and eight months, and it is made more difficult by the poor access to the interior of the boat at this time. Since the design of the 688Is makes no allowance for hard patches—points on the hull designed to be cut open—everything has to fit down the hatches leading into the interior of the sub.

TESTING/SHAKEDOWN

From the Navy's point of view, the new boat really comes to life when the reactor is powered up, or made "critical," for the first time. Prior to this, the reactor fuel elements have been loaded and a series of mechanical and electrical tests made. Before the reactor is allowed to go critical, every element of the propulsion system will have been tested under real-world conditions for a substantial period of time. During a final test (known as a Reactor Safeguard Examination), which is supervised by personnel from DNR and certified personally by the DNR himself, the crew is tested to affirm that they meet the standards set down over forty years ago by Admiral Rickover when the *Nautilus* first made ready to go to sea. And for the rest of the boat's service life, a DNR team will periodically be sent down to the boat for a continuing series of Operational Reactor Safeguard Examinations (ORSEs).

By this time the precommissioning crew has grown to the point that they can take the boat out for her initial sea, or Alfa, trials, in which a mixed Navy/EB crew will take the boat out into the Atlantic for a series of test runs. These tests are always carefully monitored and escorted, and throughout the history of the nuclear propulsion program, the three DNRs (Admiral Rickover, Admiral McKee, and Admiral DeMars) have each embarked on every new nuclear submarine to personally supervise the first sea period of the Alfa trials themselves. This personal accountability and responsibility on the part of all three DNRs, as well as their perfect safety record, has gone a long way in building confidence with the public, the Congress, and the administration in the U.S. Navy's ability to safely and successfully utilize nuclear power at sea.

Commissioning: Into the Fleet

When EB has finished building the boat to the contract specifications, it is time to finish training the crew and turning the boat into a warship. This process takes several more months. It includes weapons and tactical training, emergency procedures drills, navigation training, and actual weapons firings at the Atlantic undersea test and evaluation center (AUTEC) range down in the Bahamas. Located in the waters off Andros Island, this is an instrumented range where submarines and their crews can practice the process of operating their boat and learning to "fight" it. Somewhere during this process, the boat and her crew pass the point where they become one great war machine.

Almost six years after the contract was first signed, the final step in the process takes place. Once the Navy has determined that the boat is in all ways ready to enter the fleet, a commissioning date is scheduled, with the ceremony to be held either in Groton or Norfolk.

On this day the boat's name becomes official, the crew of "plank owners" (the original crew at the time of commissioning) is set, and the PCU submarine becomes a U.S. Navy submarine. Usually, high-ranking Navy and political figures give speeches, the commissioning captain gets to speak a few words about what this day means to him and the crew, and then, at a special moment in the ceremony, the commissioning pennant is broken out and the crew, adorned in their best Navy whites, rushes aboard and mans the boat for the first time in her official Navy career.

At this point the boat actually enters service with the fleet. But if the crew think they have seen the last of the builder's yard, they are mistaken. After the boat goes on its initial shakedown cruise, it is sent back to the yard for what is known as the Post Shakedown Availability (PSA) period. This involves taking the boat back to the yard and fitting all of the new equipment modifications that have evolved since the initial contract was signed. In addition, any warranty repairs that have become necessary will be done at this time. Following the PSA period, it will be time to head out to her new home port and the first real missions for the fleet. There probably will be only one or two of these before the CO gets word his relief is being sent. And when the commissioning captain leaves the boat, she really does belong to the fleet and the string of men who will command and sail her.

HOME BASES[1]

Once a boat has been commissioned into the fleet, it will be assigned to duty at one of the submarine bases scattered throughout the United States. These bases have the job of providing administrative and maintenance support to a boat, as well as providing housing and sustenance to her crew. Their facilities range from the ultramodern Trident facilities at Bangor, Washington, and Kings Bay, Georgia,

---

[1] As this book goes to press, massive cuts in the structure of the submarine force are being planned. These descriptions of bases and organizations are current as of March 1993.

to the turn-of-the-century New England charms of Groton, Connecticut. For the crews of the boats, these places mean home and family. Let's look at them.

Pacific Fleet

Out in the Pacific are a number of bases supporting nuclear submarine operations. These include Pearl Harbor, Hawaii; Ballast Point in San Diego, California; and Bangor, Washington. The most modern of these is the huge base at Bangor, designed to support operations of the Ohio-class SSBNs and their Trident missiles. It is located on Washington's Puget Sound, nestled into the trees of Kitsap Peninsula. Built in the 1970s specifically to support Trident operations, this is a huge facility with room to support a squadron of eight Ohio-class submarines. Currently this is Submarine Squadron (SUBRON) 17. Those who have had the pleasure to serve at Bangor have often called it one of the most comfortable and modern duty stations in the entire U.S. Navy. Also located at Bangor is Submarine Group (SUBGRU) 9. It supervises all of the submarine activities in the Pacific Northwest, including the permanent facilities for basing, overhaul, and rework at Bremerton, Washington. Technically, SUBRON 17 at Bangor is also subordinate to SUBGRU 9.

Down in San Diego is the sub base at Ballast Point. While the permanent facilities at this location are not as developed as other bases (it is literally carved into the side of Point Loma), it is located adjacent to the immense naval facilities in San Diego, and considered by the sub crews and their families a *great* place to be based.

Though the permanent facilities at Ballast Point are not as well developed as Bangor and some of the other bases, it has an amazing array of submarine tenders, floating dry-docks, and other support ships to provide infrastructure for the many boats and submersibles based there. The major submarine organization located at Ballast Point is SUBGRU 5, which has a number of subordinate units in addition to several attached SSNs and a tender. The first is Submarine Development Group (SUBDEVGRU) 1, which is equipped with several tenders and a rescue ship, as well as two research submersibles and the two DSRV rescue submarines. Also subordinate to SUBGRU 5 are SUBRON 3, with nine SSNs and a tender, as well as SUBRON 11, with seven SSNs and a tender.

Farther out in the Pacific is the submarine base at Pearl Harbor. Most of the facilities at Pearl Harbor date back to World War II, when the Pacific submarine force underwent a huge expansion to support the offensive operations against Japan. Today the base is still vital to submarine operations in the Pacific. The headquarters organization for the Pacific fleet, Commander, Submarine Force, U.S. Pacific Fleet (COMSUBPAC) is based here with a tender forward deployed at Guam. Subordinate to COMSUBPAC at Pearl Harbor are SUBRON 1 with eight SSNs and SUBRON 7 with ten SSNs. This large concentration of subs is designed to support U.S. Navy operations in the western Pacific, and boats from Pearl Harbor will frequently be assigned to support carrier groups as they rotate through the Pacific and the Indian Ocean.

Atlantic Fleet

The deepest roots of the U.S. submarine forces are back in the Atlantic. Here is where the boats are built and tested, and where most of the institutional infrastructure exists. This is also where the deepest cuts have occurred, and will probably continue to be made in the months and years to come. The winning of the Cold War has not been kind to the submarine force in the Atlantic fleet, and already one major base at Holy Loch, Scotland, with its assigned SUBRON 14 (nine SSBNs and a tender) has been completely closed down. As the submarine force continues to draw down, it is sometimes ironic to think that the Atlantic SSN/SSBN force, which did so much to keep the peace and win the Cold War, will be decimated by the victory they were so helpful in forging.

The headquarters for Commander, Submarine Force, U.S. Atlantic Fleet (COMSUBLANT) is located at the sprawling U.S. naval facility in Norfolk, Virginia. From here COMSUBLANT controls the largest force of SSNs and SSBNs in the U.S. Navy, at a number of different facilities. Farthest from home are SUBGRU 8 and SUBRON 22 (one submarine tender) based at La Maddalena, Sardinia. Though they do not have any submarines directly attached, these two units directly support the very active U.S. submarine operations in the Mediterranean Sea.

Closer to home, the Atlantic SSBN force is controlled by SUBGRU 10 at Kings Bay, Georgia. This includes SUBRON 16 with the last of the Trident I/C4-equipped Lafayette-class boats. Also under SUBGRU 10 at Kings Bay is SUBRON 20, with a force of five or six Ohio-class SSBNs and their Trident missiles. Essentially duplicating the facilities at Bangor, Washington, Kings Bay is another of the new generation of sub bases developed in the late 1970s and early 1980s. While the permanent facilities are quite nice, saying that Kings Bay is a pork barrel base is something of an understatement. Called the "Jimmy Carter memorial submarine base" by many people in the submarine force, it is something of a concession to the power of the State of Georgia, especially to Senator Sam Nunn and former president Jimmy Carter.

The other major facility on the Atlantic coast is the sub base at Groton, Connecticut. Let's go there now and get to know more about "the home of the dolphins."

Groton—Home of the Dolphins

If you drive or take a train northeast from New York City, you will come eventually to the quiet seacoast town of Groton, Connecticut. Here in this little New England seaport you will find the institutional womb of the U.S. submarine force, the U.S. Submarine Base. Within a few miles of this base is the EB building yard, as well as the schools and facilities where virtually every U.S. submariner will, at some time or another, spend time. The most important organization based here is SUBGRU 2. Based in a handsome turn-of-the-century building on the waterfront, it is the command organization for all attack submarines on the Atlantic coast. Currently it is commanded by Rear Admiral David M. Gobel, USN. This includes SUBRON 2 with ten SSNs, two support ships, and the nuclear-powered research submarine NR-1; SUBRON 10 with five SSNs and a support vessel; and SUBDEVRON 12 with six SSNs. In addition to the Groton-based units, SUBRON 2 also controls SUBRON 4 in Charleston, South Carolina (ten SSNs and a tender), as well as SUBRON 6 (seven SSNs and a tender) and SUBRON 8 (ten SSNs and a tender) in Norfolk, Virginia.

As you stroll along the Groton waterfront—and I recommend that you have an escort—you will see almost the full range of SSNs in the U.S. Navy, from the old Permit-class boats currently undergoing decommissioning, to the newest 688I-class boats like the USS *Miami* (SSN-755). At times it is a place of bizarre contrasts, as the beauty of the New England coastline merges with the low, dark, ominous shapes of the boats. Of particular interest is the dock leading to the boats of SUBDEVRON 12. This is the unit tasked with evaluating new equipment and tactics that will be utilized by the rest of the submarine force. For example, USS *Memphis* (SSN-691) is currently evaluating the first of the nonpenetrating mast periscope systems that will probably become standard on all new submarines built by the United States.

If you walk up the hill you come to the part of the base that houses the various facilities of the Submarine School. As the primary training pipeline for virtually every U.S. submariner, it is held in special reverence by the men of the U.S. submarine force. In the sprawl of dormitory-style housing, classrooms, and other buildings are some of the most sophisticated training devices ever designed. Not only do these facilities support the Submarine School with its new officer and enlisted recruits, they also provide periodic refresher training for submarine crews when they are in port. Many of the skills taught in these trainers are called brittle or perishable, since they may be forgotten if not practiced regularly.

One whole building is devoted to ship control trainers, where officers and men can learn how to control every type of submarine in the U.S. inventory. The trainers can teach you everything from how to do "angles and dangles"—maneuvering the helmsman and planesman control consoles—to the ever-popular "emergency blow." The trainers resemble those used to teach fighter pilots, and are exact replicas of the control rooms of the subs they represent.

Another trainer that will stun the untrained observer is the "buttercup," or flooding trainer. This is essentially a huge swimming pool with a replica of a submarine machinery room inside. From a control room in the side of the trainer, instructors can teach a group of men in real-world conditions how to control flooding casualties ranging from pinpoint leaks in pipes to a huge leak, over 1,000 gallons/3,375 liters per minute, in a main seawater flange connection. The idea is to control a series of leaks around the trainer that can fill it in just a matter of minutes. The training scenarios assume the feeling of a frantic fight for survival, and the crews that take the course love it for the confidence it builds and hate it for the discomfort it generates. If they do it right, the water will be roughly up to their waists if, and when, they finally control the flooding. I should say that the water for this trainer comes from a 20,000-gallon storage tank and is *very* cold.

Of all the trainers at Groton, none is more impressive than the firefighting trainer in the new facility at Street Hall. This new facility is a positive response to the firefighting casualties incurred on the USS *Bonefish* (SS-582) and the USS *Stark* (FFG-31) during the 1980s. Where previously firefighting training was conducted inside a large sewer conduit filled with blazing diesel fuel, it is now conducted in a state-of-the-art trainer that can simulate virtually every fire situation and condition that a submarine sailor might encounter. The trainer replicates, much like the flooding trainer, an engine room on an SSN. Placed strategically around the trainer are a series of propane burners designed to simulate hydraulic oil, fuel oil, electrical, and insulation (called lagging) fires.

After the crews don Nomex jumpsuits and select breathing gear—either a hose-fed compressed air mask from the Emergency Air Breathing (EAB) system or a walkaround breathing system called an Oxygen Breathing Apparatus (OBA), which uses a chemical cartridge to generate oxygen for the user—the drills begin. With all the burners lit, the temperature climbs rapidly toward the training maximum of 145°F/67°C, and there is a decided howl from the fire.

Training instructors are constantly supervising the trainees to make sure their equipment is functioning properly and they are breathing regularly, for above 130°F/58°C, the part of the brain that makes a human breathe automatically shuts down, forcing the trainees to breathe consciously on their own. In addition, the instructors add chemically generated smoke, which can reduce visibility down to about 6 inches. It is like something out of Dante's *Inferno*, and while it is exciting to watch, even the knowledge that it is a drill cannot prevent feelings of terror.

To fight the simulated fires, the trainees are equipped with a variety of fire extinguishers, fire hoses, and a new thermal imaging device called NIFTI (Navy Infrared Thermal Imager—pronounced "nifty"). This British-built device allows a sailor to "see" a fire through the smoke by the heat signature of the fire. So sensitive is the NIFTI that a human body can be located by looking for the heat of human metabolism. The fire extinguishers are designed to fight a variety of different fires. The new AFFF extinguishers, which throw a soapy slurry, are the most popular. Finally, there are a number of fire hoses that can be used to fight the simulated fires.

All in all, the Street Hall facility is a model of high-fidelity training, and similar facilities are being built at other naval bases around the United States.

All of these trainers are *very* expensive to build, operate, and maintain; in a time of declining funding, they are naturally the targets of those who would cut the defense budget. Nevertheless, I would contend that it is better to decommission an SSN or two rather than give up the valuable training that these facilities provide to the force. For while it is tough to get the money to operate and maintain an asset like a Los Angeles–class nuclear submarine, the sub is just a mass of metal without the men qualified to operate and fight her. The facilities at Groton and other bases are a tribute to the old saying that goes, "If you think training is expensive, try ignorance!"

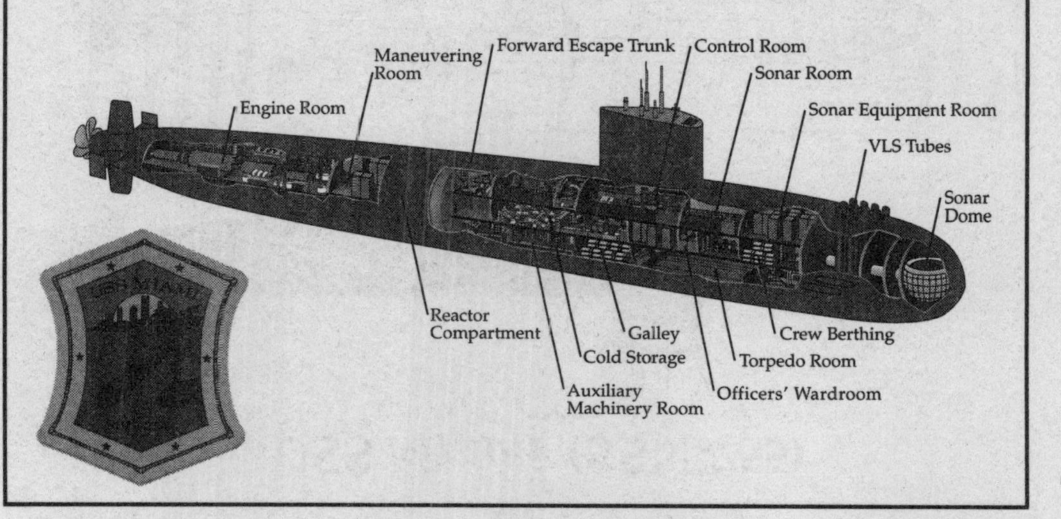

Jack Ryan Enterprises, Ltd.

USS *Miami* (SSN-755)

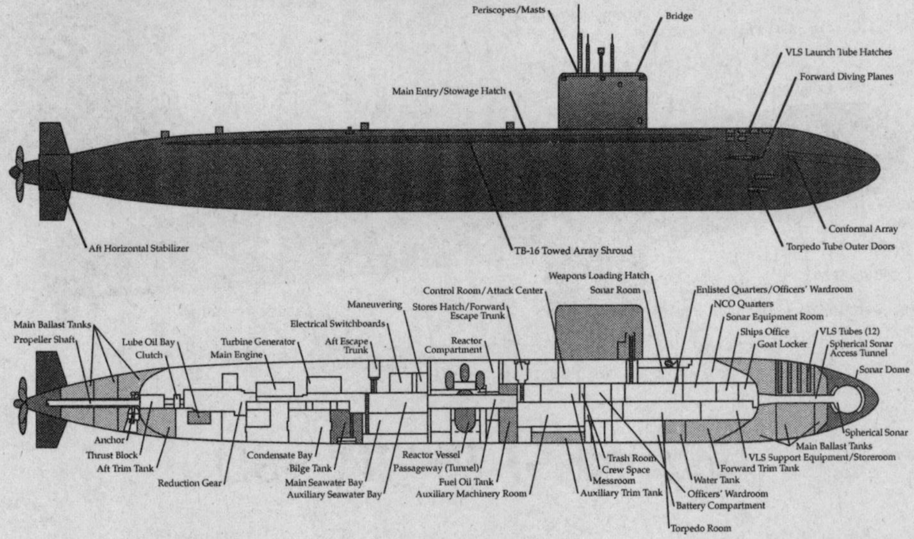

Periscopes/Masts
Bridge
VLS Launch Tube Hatches
Forward Diving Planes
Main Entry/Stowage Hatch
Aft Horizontal Stabilizer
TB-16 Towed Array Shroud
Conformal Array
Torpedo Tube Outer Doors

Weapons Loading Hatch
Control Room/Attack Center
Sonar Room
Enlisted Quarters/Officers' Wardroom
NCO Quarters
Maneuvering
Stores Hatch/Forward Escape Trunk
Sonar Equipment Room
Electrical Switchboards
Main Ballast Tanks
Ships Office
VLS Tubes (12)
Propeller Shaft
Lube Oil Bay
Turbine Generator
Aft Escape Trunk
Reactor Compartment
Goat Locker
Spherical Sonar Access Tunnel
Clutch
Main Engine
Sonar Dome
Spherical Sonar
Main Ballast Tanks
Anchor
Thrust Block
Aft Trim Tank
Condensate Bay
Reactor Vessel
Trash Room
VLS Support Equipment/Storeroom
Bilge Tank
Passageway (Tunnel)
Crew Space
Forward Trim Tank
Reduction Gear
Main Seawater Bay
Fuel Oil Tank
Messroom
Water Tank
Auxiliary Seawater Bay
Auxiliary Machinery Room
Auxiliary Trim Tank
Officers' Wardroom
Battery Compartment
Torpedo Room

Jack Ryan Enterprises, Ltd.

The Boat: A Tour of USS *Miami* (SSN-755)

The Improved SSN-688 Design

Of all the nuclear submarines designed by the United States, none has been the subject of more political infighting and controversy than the Los Angeles (SSN-688) class. The design has its roots in a series of incidents that occurred in the late 1960s, right at the time the United States was trying to decide just what kind of nuclear attack submarine (SSN) to build to replace the highly successful Sturgeon-class boats. The infighting began with the desire of then–Director of Naval Reactors (DNR) Vice Admiral Hyman G. Rickover to build a high-speed (over 35 knots)[1] submarine capable of directly supporting the fleet of aircraft carriers that represented the backbone of American seapower.

[1] The description of the development of the Los Angeles–class boats is superbly told in the book *Running Critical* by Patrick Taylor (Harper and Row, 1986).

The U.S. Navy organization charged with actually developing the specifications and design for the next generation of SSN, the Naval Sea Systems Command (Navsea), favored a design called Conform that would not be as fast as Rickover's design, but would have the advantage in areas such as habitability and quieting.

In the end, the decisive event that swung the situation in Rickover's favor was something known today as the *Enterprise* incident, which was a shock to the U.S. Navy and intelligence communities. In early 1969 the carrier USS *Enterprise* (CVN-65) and her escorts left their base in California for a war cruise to Vietnam. As she left harbor, U.S. national intelligence picked up message traffic indicating that the Soviet Union was going to dispatch a November-class SSN to intercept the carrier and her group. In an attempt to establish once and for all just how capable the first-generation Soviet SSNs were, the top battle group was provided with air cover from ASW aircraft and then told to outrun the November. It did not quite work out though, as the presumably slower Russian boat was able to match speed with the *Enterprise*. At 30 knots the game was called off. When word reached Washington, D.C., it caused rapid reassessment of just how capable the Russian SSNs really were.

Up until that point, it was assumed that the Novembers were only capable of speeds like those of the *Nautilus* and the Skates, around 20 knots. Yet here was one doing 50 percent better than that and not even trying! And what did this mean about the newer generation boats, such as the Victor I and II classes? In addition, there were mounting indications that the Soviets were working on a new class of deep-diving (over 2,000 feet/700 meters), extremely high speed (over 40 knots) SSNs.

In fact, the performance of the Novembers was due to the extreme lack of radiation shielding. Much like a hot rod that has been stripped of everything that weighs it down, the Russian boat simply did not have to haul around the reactor shielding that every other civilized nation considered essential to the good health and safety of their sailors. The November's superiority was based on a misinterpretation of the information, but there was no way to know that at the time. And Rickover was not a man to let slip an opportunity that would help justify his point of view. Through his network of Navy and congressional supporters, he pressured the Navy to kill Conform and build a class of his high-speed fleet boats. In the end he won authorization for a twelve-boat class of his fleet submarines, though to help gain critical budget authorization votes in Congress, he broke with the long-standing Navy tradition of naming submarines after sea creatures and instead named them for the home cities of the twelve congressmen who swung their votes in his favor. (Rickover is alleged to have said, "Fish don't vote!")

The first boat of the class, the *Los Angeles* (SSN-688), was to be the embodiment of his ideas of speed and power, but from the very start, it was a series of compromises. It is said that a camel is a horse designed by a committee, and the *Los Angeles* was no exception to that rule. The first problem had to do with fitting the massive S6G power plant into a hull with the dimensions needed to achieve the 35-knot speeds specified by Rickover. Quite simply, the reactor was going to come in 600 to 800 tons overweight. This meant that one or more of the key specifications of the boat—torpedo tubes/weapons load, habitability, radiated noise level, speed, sensors, or diving depth—was going to have to be reduced. The compromise was to thin the hull and limit the diving depth of the new boats to about three-fourths that of the Sturgeons and Permits (950 feet/300 meters). In addition there would be some severe compromises in habitability, forcing even more of the crew to hot bunk. As it was, there was very little reserve buoyancy (around 11 percent) and less growth potential than in any other SSN ever designed by the United States.

Once the design of the *Los Angeles* was finalized, there was the matter of selecting a prime contractor. The Navy chose the Electric Boat Division of General Dynamics Corp., despite their having submitted a bid that, in retrospect, was not capable of recouping even the costs of building the first group of twelve boats. Clearly, Electric Boat was "betting on the come"—that they could recover their lost profits from construction of boats beyond the first twelve units. Unfortunately, they did this at a time of relatively high inflation and recession in the economy, and the terms of the contract began to make it impossible for Electric Boat to break even on the first boats. Then a Navy inspection of welds found that a number of the boats had either faulty or missing welds on critical parts of the pressure hulls. This meant a number of the boats had to be completely rebuilt, further increasing costs to Electric Boat. In the end the U.S. Navy had to bail out Electric Boat and pay the costs of the overruns on the Flight I boats. This bailout caused a massive scandal that wound up costing General Dynamics the sole-source contract for the subs as well as causing the indictment of the Electric Boat yard manager on bribery charges. The Navy got the boats, but at a massive cost to the taxpayers.

On the positive side, what the Navy and taxpayers did get were the fastest, quietest, most capable SSNs ever built. On trials, the new boats proved to be all that had been hoped for them. And in 1976, when the *Los Angeles* was commissioned and sent on patrol, she clearly marked the beginning of a new era of attack boats. Part of this came from the improved sensor suite. For the first time, an integrated sonar suite was included in the design of the boat from the very start. In addition, she was among the first boats to be able to take advantage of the new family of submarine weapons, the Mk 48 torpedo and the UGM-84 Harpoon antiship missile, that were coming on line at that time. Thus what the United States got with the Flight I Los Angeles–class boats was an extremely capable camel.

This might have been the end of the Los Angeles story except for the sudden chill in the Cold War that occurred in the late 1970s. After the downturn in East-West relations, the Navy got an authorization for additional units of the Los Angeles class. And when Ronald Reagan won the presidency in 1980, the construction of additional submarines as part of the "600-ship Navy" clearly meant more Los Angeles–class boats. In these boats were to go some of the improvements that had been planned for the class early on. Starting with the USS *Providence* (SSN-719), the type designation changed to Flight II. The Flight II boats had a number of improvements, particularly in the area of weapons stowage. One of the problems with U.S. SSNs had been the limited number of weapons (around twenty-four) that could be carried in their torpedo rooms. And with the addition of Harpoon and the new family of UGM-109 Tomahawk cruise missiles (antiship and land attack versions), it was getting tougher to plan an appropriate weapons load. To get around this, a twelve-tube vertical launch system (VLS) for Tomahawk cruise missiles was added to the forward part of the boat, where room had been left for them in the original design.

Almost two dozen of the Flight II boats were built, and their cruise missile firepower proved quite useful during

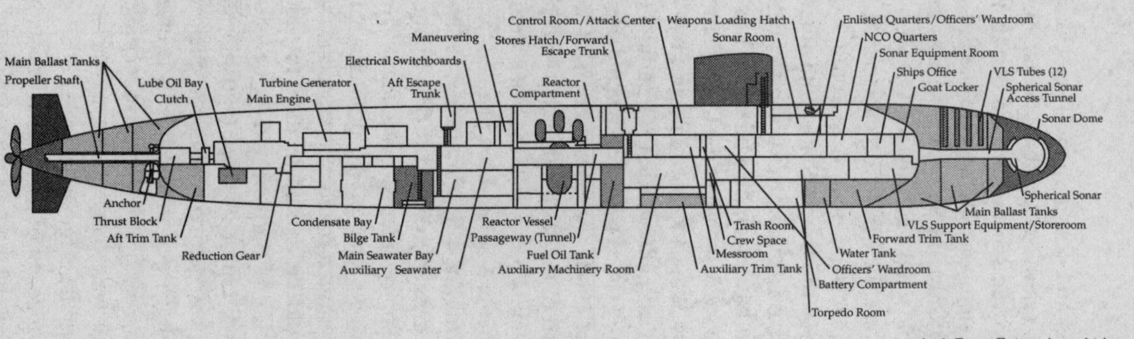

Jack Ryan Enterprises, Ltd.

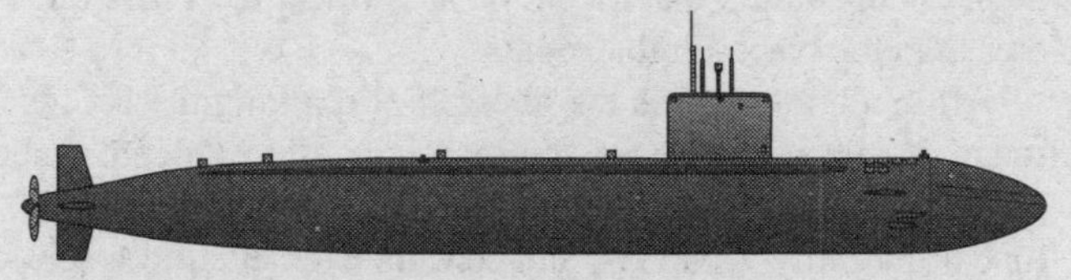

USS *Miami*, external layout. *Jack Ryan Enterprises, Ltd.*

Operation Desert Storm in 1991. The Flight IIs were also the first major group equipped with the new anechoic/decoupling coating designed to reduce the effectiveness of active sonars, as well as to reduce the noise radiated by the boat. Eventually all of the Los Angeles–class boats would be retrofitted with this coating. Another major improvement was that beginning with the Flight II boats, the S6G reactors were fitted with a new high-output reactor core. This allowed the Flight II boats to maintain their high speed (over 35 knots) despite the additional drag imposted by the new coating.

The final evolution of the Los Angeles–class boats was the version known as the Improved Los Angeles (688I). This version of the basic design would be fitted (in addition to the VLS system from the Flight II boats) with the new BSY-1 combat system. This system, which ties all of the boat's weapons and sensors together, was designed to overcome the problems associated with track and target "hand-off" between the sensor and fire control operators. In addition, the 688I was modified to support under-ice operations. This included strengthening the fairwater so that it could be used as a penetration aid through Arctic ice, as well as moving the forward dive planes from the fairwater to the hull, near the bow. Finally, the basic boat design was enhanced with a number of quieting improvements. It has been openly stated that the 688Is are almost ten times quieter than the basic Flight I boats.

All in all, the 688I is the finest SSN roaming the oceans today. While it does have shortcomings, diving depth and habitability being most notable, it still has the best single mix of mobility, weapons, and sensors ever fitted to a submarine. And while the next generation of SSNs will make up for the shortcomings of the Los Angeles class, it will be at an enormous price. In any case, the U.S. Navy had better get used to them—they have ordered a total of sixty-two boats in the class. And with the retirement of the entire Permit class, as well as planned early decommissioning of most of the Sturgeons, it is entirely likely that the year 2000 will see the U.S. Navy operating fifty to sixty Los Angeles–class boats and probably just two or three Seawolfs.

USS *Miami*: Our Guided Tour Begins

For our guided tour of a 688I, we will profile the USS *Miami* (SSN-755), the third U.S. Navy vessel to bear the name. The previous Miamis included a double-ended gunboat that fought during the Civil War, and a Cleveland-class light cruiser during World War II. The cruiser *Miami* (CL-89) earned six battle stars during her service in the Pacific during World War II, and fought in such actions as the Marianas, Leyte Gulf, Iwo Jima, and Okinawa. The current *Miami* was built at the Electric Boat Division yard of General Dynamics at Groton. She was launched November 12, 1988, and was commissioned June 30, 1990. She is assigned to SUBDEVRON 12 based at New London. She is some 362 feet long and 33 feet in diameter and has a crew of 13 officers and 120 enlisted men.

Her captain at the time of this writing is Commander Houston K. Jones, USN. He is a graduate of the U.S. Naval Academy (class of 1974), and this is his first afloat command. He is generally considered to be one of the top U.S. skippers in the sub force today, not only by his fellow officers but by the captains of the boats of the Royal Navy and other NATO nations that he has mixed it up with during various exercises. His executive officer is Lieutenant Commander Mark Wootten, USN. He is a graduate of the University of Pennsylvania (class of 1978), and is on the track to obtain a submarine command himself.

Miami is fortunate in that she is the first of the 688Is to be fitted with a complete BSY-1 combat system and all the other goodies planned for the class. The other boats of the group, starting with the USS *San Juan* (SSN-751), have less capable preproduction versions of the system and thus will have to await refits to move up to the full 688I standard. In addition, *Miami* is reported to have done 37 knots out on trials with her high-output reactor core. She is a fast, smart-looking boat with an excellent record thus far on exercises and patrols. Let's go aboard and take a look for ourselves.

Hull/Fittings

As you walk across the gangplank onto the boat the first thing that strikes you is the straight and level nature of the hull. Several things account for this. First and foremost is the fact that for most of its length, the Los Angeles–class boat is a perfect 33-foot-diameter tube of steel. This is a function of her high speed requirement. Long, narrow hulls have less drag than the teardrop-shaped hulls that can be seen on earlier U.S. or British boats. And while this does make for a faster boat, it has some adverse effects on handling during operations. In addition, it is easy to tell that *Miami* is equipped with the Mk 32 VLS system, since it is sitting level in the water. The earlier Flight I boats, because they are not equipped with the VLS, always have a pronounced "nose up" attitude when they ride on the surface.

Another thing that you immediately notice is the long shroud running down the starboard side of the hull. This is the housing for the various parts of the TB-16 passive towed array sonar. Along the shroud runs a track that allows personnel on deck to secure themselves to the hull, if surface operations are required. As you step onto the hull, you immediately notice that it seems to be made up of a series of tiles or bricks. And when you step on them, they seem to "give," much like the padding under a carpet. This is the anechoic/decoupling coating designed to defeat active sonars as well as reduce the noise emitted by the boat's internal machinery. It covers the entire hull except for the hatches, control surfaces, and sonar dome/windows.

Forward toward the bow are the twelve hatches for the VLS missile launch tubes. The outer doors or caps for the four torpedo tubes are located, two to a side, below the waterline. Along the top of the casing, aligned along the center axis of the boat, are three hatches. The one just forward of the fairwater is the weapons loading hatch. Here, using a special set of loading gear, the various weapons fired from the torpedo room are loaded. Two more hatches aft of the fairwater are set aside for the more mundane job of personnel access. Both are equipped to act as airlocks in the event that a rescue submarine needs to lock on, or as a way for swimmers to leave the boat. The aft hatch leads into the machinery spaces aft of the reactor compartment. Entry into this area is strictly controlled. The other hatch, just aft of the fairwater, is the main entry point in the forward part of the boat.

The hull is composed of a series of rings or barrel sections, welded together at the building yard. The 33-foot-diameter hull is itself approximately 3 inches thick and composed of HY-80 high-tensile steel. At each end of the 360-foot-long hull is a hemispheric end cap, which is welded onto the cylinder formed by the barrel sections. The main ballast tanks are at the forward and aft ends of the hull, with the sonar dome mounted forward and the propulsion section and its control surfaces mounted aft. In addition, smaller variable ballast tanks, which are used to maintain the trim of the boat, are located inside the hull.

Onc final thing that comes to the viewer's eye is the detail work done by the designers to minimize any type of flow noise from the hull. All of the fittings, called capstans, used to secure the boat to the pier forward of the fairwater are mounted along the centerline, so that they are already in disturbed water and will not cause any other noise on their own. No expense is spared to make the hull clean of anything that might disturb the water flow and create noise. Even the huge seven-bladed propeller, made of a special bronze alloy, is specifically designed to prevent and delay the onset of cavitation.

Sail/Fairwater

If we were to move to the top of the fairwater, we could just squeeze into the tiny bridge area. It is extremely cramped and has only the most basic of navigational aids to support getting in and out of harbor. In the past, submarine captains actually used to fight their submarines from this position. But with the advent of nuclear-powered subs, which spend most of their time underwater—*Miami* is, in fact, more stable and faster submerged than surfaced—this position has become less important.

Just behind the bridge position are the masts containing the various sensors for the boat. These include the attack and search periscopes as well as the ESM, radar, and communications masts. Some of these masts actually penetrate the hull and provide the boat with its eyes and electronic ears to the world topside. In addition, a floating antenna is reeled out from a point on the after part of the fairwater to provide *Miami* with access to the Very Low Frequency (VLF) and Extremely Low Frequency (ELF) communications channels. It trails out several thousand feet behind the boat once she has dived and stabilized. In the floor of the bridge position is a small hatch leading down some three stories into the control room. As you finally drop into the hull, you are in the port side passageway, just forward of the control room.

Control Room

Walking the few feet aft into the control room you are immediately struck by the fact that the air is clean and fresh and the room is brightly lit. And while the room is full of busy people and packed with gear, it is not really confining. One popular misconception is that if you are claustrophobic, you will not be able to live and work on a submarine—on the contrary, the very fact that over a hundred men are working, eating, and living in this confined metal tube can be reassuring.

In the middle of the control room is a raised platform with the periscopes in the middle of it. The forward part is

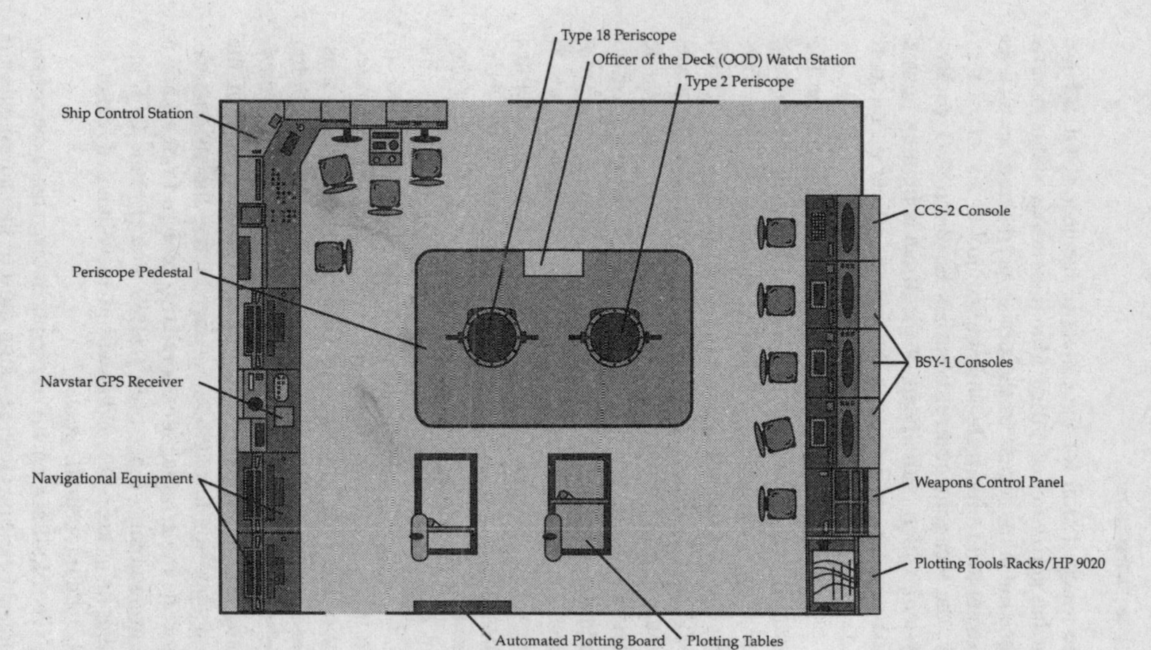

Control room, USS *Miami*. *Jack Ryan Enterprises, Ltd.*

the watch station for the officer of the deck (OOD). Here he has full view of all of *Miami*'s various status boards ahead of him, access to the periscopes behind, as well as fire control to his right and ship control to his left. These are the weapons control consoles for the BSY-1 combat system, which is the heart of the *Miami*'s fighting power. The ship control area is in the forward corner on the port side.

The navigation and plotting areas are at the rear of the compartment. Down the port side of the control room are the various navigational systems, including the new Navstar global positioning system (GPS) receiver. It is most noticeable by the gap that it sits in. Where before there was a rack of navigational equipment that took up 4 to 6 cubic feet of volume, the GPS system, which gives a three-dimensional navigational fix accurate to within 9 feet/3 meters, is a wonder taking up only about 60 cubic inches. It derives its accuracy from a series of twenty-four satellites operating in low earth orbit. The readouts show the exact latitude and longitude, as well as a number of different useful functions. So accurate is the GPS system that some U.S. Navy ship captains have been able to make blind approaches to piers in heavy fog using only GPS as a reference. The only limitation to GPS is that the *Miami* must raise a mast, such as the search periscope, to obtain a fix. To make up for this, *Miami* also has a ship's inertial navigational system (SINS) that keeps constant track of the sub's position through an advanced three-dimensional gyroscope system that senses relative motion from a known starting point. Proper use of SINS with periodic GPS updates helps keep the *Miami* within a few hundred feet of its planned track at all times.

The plotting area, aft of the periscopes, has a pair of automated plotting tables, though most of the movements are plotted by hand. Despite what one might think, most of the plotting of *Miami*'s movements is done manually by a junior officer or enlisted man, on tracing paper over a standard navigational chart. Scattered throughout the passageways are a series of upright steel boxes secured to the bulkheads. They contain several complete sets of charts which cover the entire world, as well as detailed charts for specific areas to which the *Miami* might be tasked. In addition to the navigational instruments and plots, there are a number of instruments associated with the *Miami*'s ability to work under the Arctic icepack. These include devices to obtain vertical traces of the bottom and ice floes, as well as various instruments to measure temperature and water depth.

The periscopes are mounted side by side, with the Type 2 attack scope to port and the Mk 18 search scope to the starboard. The Type 2 is a basic optical periscope with no advanced optics and only a simple daylight optical capability. The majority of the periscope work is done through the Type 18. It is the most advanced periscope currently fitted to a U.S. sub. In addition to its straight optical capability it has a low-light operating mode, which can be projected onto a number of television monitors around the boat. It is also equipped with a 70mm camera for taking periscope photos, as well as the readouts for the Electronic Support Measures (ESM) receiver mounted on top of the Type 18 mast. It also has an antenna for the GPS receiver mounted on it. This is a truly great scope, capable of almost any activity that might be asked of a periscope. The masts for the two scopes go up through the fairwater; they may be coated with a radar-absorbing material (known as RAM) to keep their radar signature down.

The ship control area, located in the forward portside corner, has three bucket seats—with seat belts—as well as room for another person to stand. Normally it is manned by two enlisted personnel who operate the diving planes and rudder (called the planesman and helmsman), and the diving officer and the chief of the watch controlling the ballast and trim. The planesman and helmsman are faced with aircraft-style control wheels, and sit facing a bank of control readouts and instruments. There is no view of the surrounding sea and even if there were, it would do little good. At depths over a few hundred feet very little light penetrates, and the sea becomes, as Jacques-Yves Cousteau calls it, "a dark and silent world."

Just behind the ship control area stands the diving officer, who is actually ordering the planesman and helmsman what to do and when. To his left is the position where the COB may sit, though others will frequently draw duty there. This is where the controls for the multitude of valves, tanks, and other equipment required to dive and surface the boat are located. Each man controls either the rudder and bow planes, or the horizontal stabilizer. Two-man control has been a hallmark of U.S. design philosophy for generations, and *Miami* is no different. For every primary system there is a backup, usually with a manual operating mode. Most noticeable of these are a pair of mushroom-shaped handles located at the top of the ballast control panel. These are the manual valves to conduct what is known as an emergency blow. In the event the boat needed to get "on the roof" in a hurry, the person at the ballast control panel would activate these two handles. These valves, which require no power of any kind, send high-pressure air directly from the air banks into the ballast tanks—when that happens, you're headed up fast. Early American SSNs did not have this feature, and this lack was felt to be a contributing cause of the loss of the *Thresher* in 1963.

Diving the boat is not the crash dive of 1950s submarine movies. In fact, it is a carefully controlled and balanced procedure that resembles a ballet danced by an elephant. First, the captain orders any personnel down from the bridge, and the closing of all hatches. Once that is done, the diving officer looks over the status board to the left of the ship handling stations to verify that all hatches and vents are sealed, and that the air banks have an appropriate reserve of air pressure. This done, the diving officer opens the vents atop each ballast tank to allow a measured amount of water into the tanks. This is just enough to make the boat slightly heavier than the surrounding water (called negatively buoyant). As this is happening, the diving officer orders the planesmen to put 10 to 15 degrees of down angle on the boat, using the bow and stern diving planes. At this stage the boat begins to settle. All told, this process normally can take from five to eight minutes.

Initially the dive will be held up when a depth of 60 feet (periscope depth) has been reached. At this point the depth will be maintained with the dive planes and the forward motion of the boat. During this time the diving officer will have the chief of the watch pump water in and out of the trim tanks to make the boat neutrally buoyant and balanced. In addition, the captain will probably order a series of checks on all of the compartments of the boat for watertight integrity, and do a check to see if any machinery is making abnormal noises, or if any objects are loose or improperly stowed. Next the captain will probably order a series of extreme diving exercises called angles and dangles, which are designed to discover if anything is still improperly stowed. The old hands take a perverse pride in being able to walk and keep a cup of coffee from spilling during high-angle dives. Now the *Miami* can get down to cruising.

Maneuvering a 6,900-ton submarine is something that is done with subtlety and a minimum of rapid action. A slow and delicate touch on the planes and rudders is required to prevent unwanted noise. If you desire to change speed, you rotate a knob called an Engine Order Telegraph, which sends an instruction back to the engine room to either increase or decrease the power to the propeller shaft. The lack of precision might surprise some people, as there are only Forward and Reverse, with choices for All Stop, One Third, Two Thirds, Full, and Flank. In spite of this, the precision that you can maneuver the boat with is amazing. In fact, the OOD can order the precise number of propeller revolutions or "turns" required to maintain any speed required.

The one problem with driving a 688I is that it tends to be slightly unstable at some depths and speed settings. This is partly a product of the 688I's hull shape, which is optimized for speed, and partly from the forward placement of the fairwater. Normally only light corrections will be necessary to keep tracking but one must be ready for any situation, including combat maneuvers, which can become downright violent.

Running underwater is, if nothing else, probably the smoothest ride that you will ever know. Once the boat is trimmed and level, there is little or no sensation of motion, and you feel as if you're walking through the basement of a building. There is, in fact, a feeling of being on very solid ground. Very reassuring, and very quiet. In fact, quiet is the name of the game in this business. When the sub is running underwater, nobody raises their voice, slams a hatch, or even drops the toilet seat hard. After a time, you become hushed and silent. So much the better.

Surfacing the boat is an exercise in itself, as there is no more vulnerable time for a submarine. Part of this is because a surfacing boat makes lots of noise: the rush of compressed air from the air flasks into the ballast tanks; the noise of the hull expanding from the decreased water pressure, called hull popping. All this noise makes the boat partially deaf and blind, so special precautions are taken. The first thing the diving officer does is to have the planesmen at the ship control stations bring her to periscope depth. At this point the search periscope will be raised to do a visual check for any surface vessels, as well as sonar listening for any surface or subsurface contacts. Once the captain is confident that all is clear topside, he will order the diving officer to blow compressed air from the air flasks into the ballast tanks to give the boat a slightly upward, or positive, buoyancy. Within several minutes the boat will surface, and the captain will establish a bridge watch up on the fairwater.

Once on the surface, you immediately notice the rolling of the boat in the surface swells. It is an ironic truth that the same hull design that provides such a smooth ride in the depths of the ocean rolls rather drunkenly in a mild surface swell. While it is not particularly uncomfortable, when compared to the amazing stability of the boat at depth, the difference seems enormous. While running on the surface, it is essential that the bridge watch maintain a constant lookout for any surface vessels. Since a submarine is as hard to see as it is, submariners are always concerned about being run over by a rogue supertanker or liner, and are cautious to avoid fishing vessels, especially those using drift nets.

Communications/Electronic Warfare Spaces

The communications shack is located forward of the control room along the port side passageway, and is notable for the security warnings posted on the door. It is incredibly vital to the operations of *Miami*. Packed into that tiny space is all of the radio transmission and cryptographic gear that is required to send and receive messages, ranging from operational combat orders to personal "familygrams."

The radio equipment covers a broad spectrum of frequency ranges from ultra-high frequency (UHF), high frequency (HF), very low frequency (VLF), and extremely low frequency (ELF). In addition, there is equipment designed to allow the *Miami* to contact communications satellites, as well as underwater telephone equipment commonly known as Gertrude. Most of the radio equipment is tied to sophisticated encryption gear (called crypto) designed to make it impossible for anyone but an American to read the message traffic.

This particular point has not always been so secure, as the discovery of the Walker family spy ring showed in 1985. For over fifteen years, a Navy petty officer, along with his family members and a friend, helped the Soviet Union acquire the keys to the various crypto systems used by the United States. This meant that the Soviets had access to virtually all our major crypto systems from 1969 to 1985, when the ring was finally apprehended. Since that time the National Security Agency, which is charged with the design and security of crypto systems, has apparently rebuilt the U.S. family of encryption systems and allegedly changed the procedures that allowed John Walker and his family to put so much of our national security at risk.

The most interesting of these systems are the ELF and VLF systems, which are mainly used as command and control systems for submarines. Their special property is that the signals from ELF and VLF systems can penetrate the water to be picked up by the antenna trailed from the port side of the fairwater. More often than not, because of their relatively low rate of transmission (ELF works at about one letter character every fifteen to thirty seconds; VLF is fast enough for teletype communications), they are used to cue submerged submarines to come to periscope depth, and poke one of their communications masts up to get a signal from a satellite or UHF channel.

It is standard on submarines to minimize any actual transmission from their radio systems. Always looming over the submarine force are the memories of what the Allied ASW forces were able to do to the U-boats in World War II, because of their knowledge of the German Enigma cipher system. The penetrations of U.S. systems by the Walker spy ring have only reinforced the belief that transmitting with a radio is an invitation to a funeral. Thus it is only occasionally when they are close to a potential enemy that they will send messages. To a submariner, only silence is a friend. Any noise, acoustic or electronic, is an enemy.

Another method of communicating with the outside world is for the boat to eject a SLOT (Submarine-Launched One-Way Transmitter) buoy from its forward 3-inch signal ejector launcher. Located in a small compartment forward that doubles as the ship's pharmacy, it resembles a tiny torpedo tube. The first step is to record a message, such as a contact report, on the buoy's recorder. The buoy is then fired into the water, where it waits a period of time, say thirty minutes to a couple of hours, then sends out a high-speed burst transmission that can be picked up on a special satellite communications channel.

In addition to launching SLOT buoys, the 3-inch ejector can be used to launch bathythermographs to monitor thermal layers in the water, as well as several types of decoys such as noisemakers and bubble generators. A second 3-inch ejector is aft in the engineering spaces, and both units can be controlled and fired from a panel in the control room.

Keeping track of the electronic noises an SSN encounters is the job of *Miami*'s Electronic Support Measures (ESM) suite. Technically the suite is made up of a radar and electronic signal receiver known as WLR-8 (V). This is used to monitor the radar and radio emissions in operational areas. In addition, the *Miami* is equipped with a BPS-15 surface search radar to assist in ship handling and navigation. All these systems have their antennas mounted on retractable masts, which can be raised while the boat is at periscope depth.

AN/BSY-1 Combat System

At the very heart of the *Miami*'s combat power is the new BSY-1 (pronounced "busy one") submarine combat system. All the sensor, fire control, and weapons systems of the Flight I and II Los Angeles–class boats, as well as a few new items, are tied together into a single system controlled by a battery of UYK-series computers running almost 1.1 million lines of Ada (the defense department's systems programming language) computer code. Developed by IBM, with Hughes, Raytheon, and Rockwell as subcontractors, BSY-1 represents the first use of what is known as distributed processor architecture. All of it is tied together by a data highway known as a data bus, which is becoming something of a standard on weapons systems such as the F-18 Hornet fighter/bomber and the Patriot surface-to-missile system.

This means that instead of having one large computer running all the sensor and combat functions, a central computer hands out processing assignments to other computers running code designed to handle a specific job like acoustic processing or cruise missile mission planning. In

The placement of the *Miami*'s forward sonar arrays. *Jack Ryan Enterprises, Ltd.*

this way the distributed system actually runs faster than a larger single computer would. It also makes the BSY-1 system easier to upgrade and better able to operate in a degraded or damaged condition.

Other than the racks of UYK-7, UYK-43, and UYK-44 computers buried in the computer compartments, the most visible signs of the BSY-1 system are the consoles in the sonar room, forward of the control room, along the starboard passageway. Here four manned sonar consoles provide the *Miami* with her ears to the underwater world. Into these consoles the BSY-1 system feeds information from the various sonar systems. The *Miami*'s main sonar system, almost identical to the BQQ-5D system on earlier Los Angeles–class boats, is actually a collection of many different sonar systems, including:

- The spherical sonar array, located in the bow. The large sphere (15-foot diameter) has both active (echo ranging) and passive (listening) modes, and is currently one of the most powerful active sonars (over 75,000 watts of radiated power) afloat anywhere in the world.

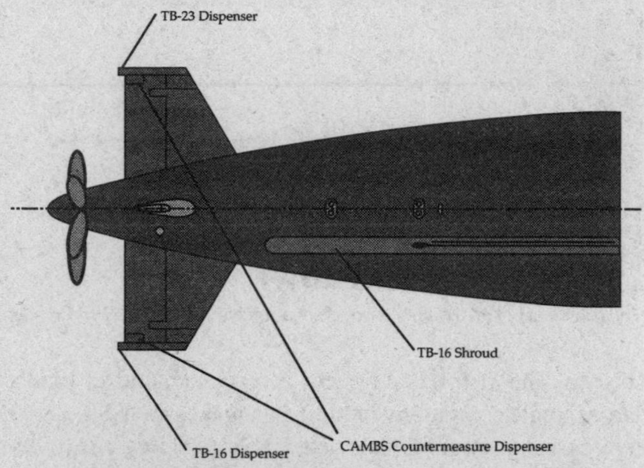

Mounting of the *Miami*'s towed sonar arrays. *Jack Ryan Enterprises, Ltd.*

- The conformal array is a low-frequency passive sonar array mounted around the bow.
- The high-frequency array is an upgrade to the spherical array, allowing it to generate the advanced waveforms that make the active modes of the BSY-1 so effective. It also incorporates an under-ice and mine detection capability from an array in the fairwater.
- The TB-16D is the basic towed array, which is fed from the tubular shroud on the starboard side of the hull. It is passive system, designed to provide medium-range detection of low-frequency noise. It is fed from a large reel in the forward part of the boat and played out from a tube in the starboard horizontal stabilizer. It has a 2,600-foot cable that is 3.5 inches/89mm thick, with the receiving hy-

drophones in a 240-foot-long array at the end of the cable.

- The TB-23 is the new passive "thin line" towed array associated with the BSY-1 system. Its smaller diameter (1.1 inches/28mm) means that the hydrophone array can be longer (approximately 960 feet), and it can be farther away from the noise of the towing submarine. The TB-23 is specifically designed to detect very low frequency noise at very long ranges. It is stowed on a reel in the aft and fed from a receiver in the port horizontal stabilizer.
- The WLR-9 is the acoustic intercept receiver designed to alert the crew that an active sonar is being used, such as large active sonar arrays or sonar on incoming weapons.

Associated with all these systems is a series of signal processors and other equipment, which translate the sounds emitted and collected by the various sonar systems into the data displayed on the sonar consoles. The four BSY-1 sonar consoles are usually configured to have three of them looking at particular elements of the BQQ-5D sonar sensors while the fourth is used by the sonar watch supervisor. There also is a sonar spectrum analyzer available at a workstation in the forward end of the compartment. Each console has a pair of multifunction displays, which can be configured quickly by the operator for the particular sensor and mode of interest. For example, one sonar technician might be looking at the broadband noise being collected from one of the towed arrays. Another might be watching for broadband contacts on the spherical array.

What the sonar technician actually sees is a rather odd-looking display called a waterfall. It looks like a green television screen full of snow or "noise." The top of the display shows the bearing of a particular noise source or frequency being detected. The vertical scale shows that noise or frequency over time. The sonar technician is looking for something that stands out from the random pattern of background noise being displayed. Usually the sound contact appears as a solid line on the display screen. And this is where the hunt begins.

The technician reports the contact to the sonar watch supervisor and begins the process of classification and identification. The supervisor alerts the officer of the deck that a new sonar contact, called "Sierra Ten," for example (contacts are numbered progressively), has been detected and that the sonar team is working it. The conventions for naming contacts are:

- Sierra — a sonar contact
- Victor — a visual contact
- Romeo — a radar contact
- Mike — a contact combining one or more signals from different sensors

What is important now is patience and concentration. And much like my character Jonesy, these technicians pursue just as much an art as a science. As soon as the first sound line has established that a contact exists, the other technicians assist in the classification. Despite all that has been written before, there is no automatic classification mode in the boat's computers—one of the *Miami* sonar technicians has proudly said, "We still do it ourselves."

Sometimes a frequency line is known to be unique to a particular power plant of a particular ship or submarine

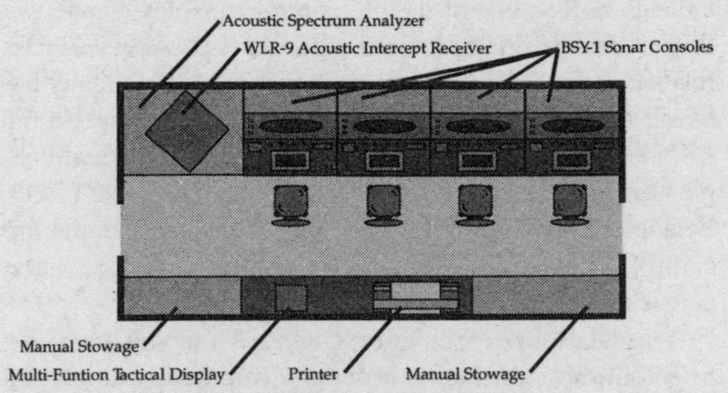

Sonar room, USS *Miami*. *Jack Ryan Enterprises, Ltd.*

class. Other times, the effort to classify the target may require the technician to listen through headphones to try and make out what the signal on a particular bearing is. They can listen to tonal signals to determine whether the source is a surface ship or submarine. Each of the different sonars in the BSY-1 suite has its optimum frequency band, and if another sensor might be better at getting data on a particular signal, the technician is fully empowered to ask the officer of the deck to alter course to bring that sensor to bear. During this time the sonar watch team are the eyes and ears of the boat, and every other man aboard knows that his safety may depend on just how good the operators in the sonar room really are. There are set procedures to help guide the sonar technicians, but in the end it comes down to the individual skills of the technicians doing what must be a mind-bending job.

The sonar supervisor reports the best estimate of what and where the source is, and whether it could be a threat or not, to the officer of the deck (OOD). The OOD stations the fire control team to begin the localization/tracking process. This is a dual process utilizing both the manual plotting table as well as one of the fire control consoles. On *Miami* this process is different from the older Los Angeles–class boats in that all the information is passed automatically between the sonar room and the fire control console via the BSY-1 system network. At this point the tracking team begins the process known as Target Motion Analysis (TMA). Besides identifying the contact, the TMA provides the fire control team with a usable fire control solution, target course and speed, and a reliable range.

This takes time—sometimes, a lot. While you are trying to get all the information necessary to possibly shoot at a target, you must yourself remain undetected. Much of the data for the TMA process comes from the bearing rate, which is how fast the bearing of a target is changing, and monitoring the Doppler, which reveals whether a target is coming nearer or moving away; this is called the range rate. While the BSY-1 is helping the fire control team do its job, the manual plot team, assisted by a specially programmed Hewlett Packard 9020 desktop computer, is also working on its own TMA/range analysis. This little desktop computer has a program library that helps the manual plot team with the more intensive calculations and generates what can only be called instant ranges to the target. All the while the manual and automatic tracking solutions are checked, and data is crossfed between them. During the TMA process the boat would probably maneuver in a zigzag pattern to help the sonar crew establish better range and bearing rates for the TMA plots.

Some nations have chosen to eliminate the dual TMA process and depend only on an automatic system. But this can lead to ranging errors in critical situations, so the United States Navy continues to use manual plots and automatic systems just to be sure. Recently *Miami* ran an exercise against a diesel boat belonging to one of our NATO allies. Apparently, because *Miami* had a small acoustic fault (called a sound short), the opposing sub thought the boat was much closer than it actually was: the automatic fire control system calculated the range to *Miami* at around 6,000 yards when, in fact, it was over 40,000 yards. And when the diesel boat fired at what it thought was a nearby U.S. boat, all it did was expose itself to attack by the *Miami*. Needless to say, Commander Jones made his "opponent" pay dearly for his error.

The TMA process is continued until the commanding officer believes the tracking party has a good enough picture of the situation at hand. Every contact has to have a reliable TMA solution and must be currently tracked. Here lies the real value of the BSY-1 system. For while the earlier Los Angeles–class boats could keep track of only a few targets at one time, the BSY-1 can handle many more. And once the system has a good track running, it has a great ability to hold and maintain the quality of the tracks.

Eventually the target track(s) will be good enough to fire on, if that is the desired intention, and the time has come to set up a weapon for firing. The fire control technician begins the process by inputting the necessary presets into the chosen weapon. If it is a Mark 48, Harpoon, or Tomahawk antiship missile (TASM), this can be accomplished entirely at the BSY-1 console. Should a Tomahawk land attack missile (TLAM) need to be programmed, this is accomplished at the adjoining Command and Control System (CCS-2) console. For now, though, we will concentrate on the weapons programmed on the BSY-1 console.

If, for example, the desire is to launch an antiship missile, the technician must have a decent estimate of target course, speed, and range. It is also critical to know whether there is any neutral shipping traffic in the area. The technician programs in the route to the target, as well as any waypoints necessary to route the missile around neutral shipping traffic that might be in the way. In addition, the technician programs a search pattern for the seeker head of the missile to lock in. This mission plan can be loaded into any number of missiles, which are then fired from the weapons control console located to the right of the fire control consoles.

The process for firing torpedoes is somewhat more dynamic than that for missiles. First the fire control technician develops a fire control solution through a process called "stacking the dots." The screen where this is accomplished displays the target bearing versus time, similar to that back in the sonar room. On this display the target bearing is shown over a period of time as a series of dots. The technician fine-tunes the solution by adjusting the estimates of the target's range, course, and speed until the display shows a straight column of dots stacked on the display. After several minutes of work and possibly a couple of maneuvers to verify the accuracy of the solution, it is now time to shoot.

Despite what some computer games would have you believe, there are no joysticks for the fire control technicians to "fly" the torpedo onto the target. Instead, the technician changes the weapon presets on a screen that looks like a shopping list of parameters such as the seeker activation point (called "enable run"), search depth, and which seeker head mode the weapon is to be fired in. Also, the BSY-1 has several different operating modes, including a "snapshot" mode for fast-moving tactical situations that require the *Miami* to react quickly. Let's assume that the fire control technician has been ordered to set up a pair of Mk 48 ADCAP torpedoes for a shot at a submarine. He selects the desired target track and allows the BSY-1 to input the weapon presets to the list.

At any time, he can override or alter the presets to suit the tactical situation. For example, the ADCAP has modes to avoid making circular runs and attacking the firing sub accidentally, as well as the ability to preset a three-dimensional search zone for the weapons to search in, but not go outside. Once the weapons have been loaded with the required data, they can be fired by the weapons officer at the order of the captain. With the weapons now in the water, a junior officer calls up the weapons display on his console and monitors the torpedoes' status.

One of the nice features of the BSY-1/ADCAP combination is that the technician can "swim" the torpedoes out onto the target and use the seeker heads as offboard sensors to fine-tune the firing solution. This is made possible by the data link wire that the weapons trail out behind them, which is connected to the torpedo tubes of the *Miami*. This means that if the technician sees the target move out of the selected area, or do something tactically different from what he thought it would do, he can quickly change the necessary presets right from his weapon control menu.

When the ADCAPs finally acquire the target, the process becomes completely automatic, with the operator's help required only if a torpedo malfunctions. The logic in the guidance systems of the ADCAPs is very good, though if anything goes wrong the fire control technicians are always ready to step in on their own. Assuming that the weapons do their job, the final run to the target will be like watching a train wreck. When they hit, the sonar technician must assess the damage that has been inflicted. There may be breaking-up noises or the distinctive *crunch* of an imploding pressure hull. In any case, the tracking teams are now ready to start again, a never-ending task while on patrol.

One thing we haven't mentioned yet is just why the *Miami* has an active sonar mode when so many great things can be accomplished just by listening passively. For almost thirty years, going active with a sonar has meant giving up the tactical advantage. The simple truth is that while using an active sonar does alert a potential enemy to your presence, it does have some significant advantages. The latest nuclear boats produced by the former Soviet Union/Commonwealth of Independent States are almost as good acoustically as a Flight 1 Los Angeles. This means that finding them passively is going to be extremely difficult. And the current generation of diesel boats, when running on their batteries, are just a little worse, being *very* quiet targets to any passive sonar system in existence. Using an active sonar can overcome some of these problems at relatively short ranges, and has tactical benefits in some situations, especially in verifying ranges before shooting. Unfortunately, an active sonar can be heard at least five times farther than the sonar can detect a target.

The active sonar mode of the sphere sonar is incredibly powerful and can cause steam bubbles to form on the outer surfaces of the sonar dome. The spherical array does give accurate ranges and bearings, providing excellent fire control solutions in the process. In addition, it has the ability to form its sound signals into beams that are focused instead of just radiating in all directions. This means that only the target boat will know it is being "pinged," and other boats in the area will not. In the sort of close-range "knife fights" that may develop between the quieter boats inhabiting the oceans today, going active may just be a good thing to do.

This is a rough picture of how the BSY-1 system and her operators work together. Many other elements go into the process, but I hope this has given you a feel for how the operators would use BSY-1 to fight the boat. If you think it

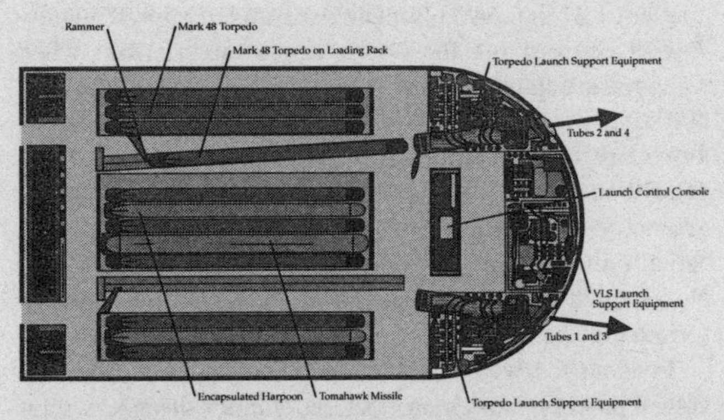

Torpedo room, USS *Miami*. *Jack Ryan Enterprises, Ltd.*

seems like a huge game of blindman's buff, you are right on target, for it is said that in the land of the blind, the one-eyed man is king. In the dark realm of the world's oceans, the *Miami* with its BSY-1 combat system is the king with the biggest eye.

Torpedo Room

When you wander down a couple of flights of stairs and move forward, you eventually wind up in the torpedo room. Here you are struck by the feeling of being in the very bowels of the *Miami*. Three sets of two-high racks allow for the stowage of twenty-two weapons, and four more are kept in the tubes. Usually, however, one or two of the rack spaces or tubes are left empty, to facilitate movement of the weapons and allow maintenance. Between the center and side racks are sets of loading and ramming gear. Go forward down the aisles between the racks and you will find the torpedo tubes. These have an internal diameter of 21 inches/533mm and are angled approximately 7 to 8 degrees off the boat's centerline, so that when weapons are launched, they clear the bow with its big active sonar dome. One unique design aspect is the ability to move any weapon from any position in the racks to a torpedo tube or any other position on the racks. While the geometry of such a move is somewhat complicated, the actual movement of the weapons resembles a child's puzzle in which eight pieces are moved through nine spaces to form a picture.

Loading the weapons into the boat itself is a rather involved process, though one that the *Miami*'s designers actually thought out pretty well. Just forward of the fairwater is the weapons loading hatch; through here the weapons are brought on board. The first step in the process is to open this hatch and unstow the loading gear, which is cleverly composed of sections of the flooring structure from the second and third decks of the boat. The second-deck flooring becomes a loading rack that is hoisted up on deck to receive the weapons from the loading crane alongside. A section of the third deck serves as the transit rack, which spans the gap left by taking up the floor structure. Thus while loading is taking place, a gap like a canyon runs down the middle of the boat to the torpedo room.

The actual weapons-loading process is quite rapid once the gear is assembled. The weapon is swung over on a crane from the dock or tender and gently lowered into the loading rack. Once it is aligned, the loading rack is rotated up about 45 degrees, and the weapon is winched down on a chain-powered hoist. When the weapon has completed its nearly 50-foot journey, the transit rack is swung back to the horizontal, and the weapon is laid into the waiting skids on the torpedo room racks. At this point it is secured to the skids and moved over so that another weapon can be loaded. In all, the boat can be completely loaded, including setting up and striking the loading gear, within twelve hours, all with minimum support from a tender or dock crew. Afterwards, when the deck structures have been put back in place, you would never know this is the path the weapons take to the torpedo room.

Loading a torpedo, while straightforward, is anything but simple. The first step is to move a weapon from the storage rack onto one of the loading trays. This requires a bit of brute force (Mk 48s weigh about 3,400 lb/1,545 kg) as well as some precision; even in this day and age, human brawn is still useful. Once the weapon is loaded onto the tray, the inner door (called the breech door) to the chosen torpedo tube is opened and a quick inspection is conducted. If another weapon has just been fired, the crew may need to remove a wire dispenser and/or some guidance wire (if it is a Mark 48 torpedo), or to check for wear on the tube. This little process, known as diving the tube, is a job best handled by those with narrow shoulders and long arms.

Once this is done, the loading ram carefully moves the weapon into the tube. At this point one of the torpedoman mates (TMs) connects the data transmission link, called an "A" cable, from the back of the weapon (all U.S. submarine-launched weapons are equipped with such connections), attaches the guidance wire (if it is a Mk 48), and seals the breech door. Once the hatch is closed, the technicians check to make sure all the connections and seals are properly set, then hang on the tube a small sign: WARSHOT LOADED. One of the nice features on the 688I/BSY-1 boats is that once a tube is loaded, it automatically can tell what kind of weapon is loaded. On several control panels and status boards around the boat, the change in the tube's status to Loaded and what it is loaded with are noted and marked.

Once a decision to launch a weapon has been made (this always requires a look at the mission orders and the standing rules of engagement), then the technicians at the BSY-1 firing control panels up in the control room power on the weapon to warm it up. Then the fire control technician assigned to control the weapon loads targeting and other data into the weapon's memory system. In the case of a Mk 48, this includes speed settings and seeker head mode. For a guided missile like a Tomahawk, it involves loading a complete mission flight profile. Once this is done, the weapon is ready to be fired.

The process of firing a weapon from a torpedo tube is probably one of the most well tested procedures on the entire boat; it dates back many decades. With the weapon warm and ready to fire, the order is given, "Make the tube ready in all respects!" This is not done lightly, for this is the first of a number of actions that radiates a great deal of noise into the surrounding water. Once the tube is flooded, the outer door or cap is opened, and the tube is ready to launch the weapon. The commanding officer gives the command, "Firing point procedures," when the other necessary steps (such as sealing the breech door) have already been completed.

At this point the captain issues the firing command, "Match bearings and *shoot*!" When the order to fire is given, the weapons officer at the BSY-1 launch control panel presses the firing button, and the firing sequence begins. The firing command directs high-pressure air from the air banks onto a piston. The air forces the piston to move along the piston shaft, forcing water out of another tube and through a slide valve in the rear of the torpedo tube, thereby forming a water ram that ejects the weapon out into the sea at something like four to six times the force of gravity.

What happens next depends on which weapon has been fired. If it is a guided missile, then the outer door can be closed, and the tube is drained and made ready for reloading. If the weapon is a Mark 48, then the decision will probably be made to leave the outer door open. This is because the Mark 48 trails a guidance wire behind it, which allows the boat to guide the torpedo as it runs up to ten miles from the launching point. At any time, though, the wire can be cut. If the sub is traveling too fast, or makes too sharp a turn, then the water flow may break the wire. In any case, until the need for the guidance wire is gone, the tube must stay in use.

Vertical Launch System (VLS)

One of the weaknesses of all U.S. attack submarines since the Permit-class boats hit the water has been the shortage of space for torpedo tubes and weapons stowage. For over thirty years, U.S. attack boats have always had four 21-inch/533mm torpedo tubes to deliver their weapons, and about twenty-two stowage positions to hold them inside the boat. This was not much of a problem so long as all that the boats had to fire were heavy torpedoes and the occasional SUBROC. But beginning in the late 1970s with the introduction of the UGM-84 Harpoon antishipping missile, and the early 1980s with the UGM-109 Tomahawk missile series, this began to pose a real problem for submarine planners and skippers.

For example, say a U.S. sub skipper wants to shoot Harpoon missiles at a surface warship. Submariners traditionally prefer to keep at least one torpedo in a tube as a just-in-case weapon, much as a police officer keeps a hideout weapon in an ankle holster. This means the maximum salvo size that can be fired at the target ship is three Harpoons. This might be fine, but against a target like a Kirov-class battle cruiser with all its antimissile systems, those three missiles will be soaked up like water into a sponge; the weapons will be wasted, and the target will be alerted to the presence of the sub. What clearly is needed is a way to stow more weapons on the boat and fire more of them at one time.

The designers of the Los Angeles–class boats anticipated this, because both the designs for Harpoon and Tomahawk were known at that time. Space was left in the forward ballast tank for twelve Vertical Launch System (VLS) tubes, each capable of storing and launching a Tomahawk cruise missile. In addition, space for the associated control and hydraulic systems necessary to operate the VLS system was left in a compartment forward of the torpedo room. Thus it was possible for a Los Angeles–class boat to carry and launch twelve additional cruise missiles without affecting the weapons stowed and fired out of the boat's torpedo room. This meant an increase of 50 percent in weapons stowage and a 400 percent increase in ready firepower (when firing cruise missiles) over a non-VLS sub.

This change was not made immediately, however. Even though all the Los Angeles–class boats were capable of being fitted with the VLS system, the first boat to be so equipped was the USS *Providence* (SSN-719). And, because of budget constraints, it is quite unlikely that any of the earlier Flight I boats will ever be retrofitted with VLS missile tubes. Nevertheless, by the time the class is finished building, some thirty-one Flight II and 688I boats will have the system, providing room for some 372 Tomahawk missiles in the fleet. And that is a *lot* of firepower. By the way, it is easy to make out which boats have the VLS and which don't by whether they are level in the water (VLS equipped) or nose up (non-VLS Flight I).

The way the VLS system works is quite simple. The missile canisters are loaded vertically from a crane. Each canister contains a complete all-up Tomahawk round, ready to fire. At the top of each canister is a thin membrane of clear plastic, which keeps the missile dry and safe. This is how it stays until the time to fire. The boat comes to launch depth, usually about 60 feet, and reduces speed, say 3 to 5 knots, perhaps raising a communications mast to get additional targeting or a navigational fix from the GPS satellite constellation. Once the flight instructions have been programmed into the desired missile(s), the launch system automatically begins the firing sequence.

The system opens the missile launch tube hatch hydraulically and an explosive charge propels the missile up through the plastic membrane and into the water. After the missile travels up about 25 feet the booster rocket fires, thrusting the Tomahawk out of the water. At this point the missile tilts over, drops the burned-out booster motor, lights the turbojet engine, and heads for its preprogrammed target. Meanwhile the launch tube fills with water (helping to compensate for the lost weight of the missile), and the hydraulic hatch is closed.

The VLS system is causing a revolution in design of new weapons for submarines. It has radically increased both the firepower and stowed weapons load for the U.S.

submarine force—all at no increase in the size or displacement of the basic Los Angeles design.

Living Spaces

On the *Miami*'s second level is the bulk of the living space aboard the boat. If you stand aft near the forward escape trunk, then you walk forward, you will find the largest open area on the boat, the enlisted mess area. This place is a combination of cafeteria, schoolroom, movie theater, game room, and almost anything else that involves gathering the boat's enlisted population together. Here are six tables with bench seats on both sides so that something like forty-eight sailors at a time, about half the *Miami*'s population, can sit down at once. Along the starboard bulkhead are such cherished pieces of equipment as the soda machines (no longer do they serve the hated "Yogi" cola), milk dispenser, soft ice cream machine, and that most cherished of Navy wardroom icons, the bug juice dispenser. By the way, well-informed palates suggest that the red flavor is best, but stay away from the orange! Strangely, it also makes an excellent scouring powder for cleaning floors and heads (all that acid in it, they tell me). Back near the escape trunk is the ship's laundry. About the size of a phone booth, it handles the laundry for the entire boat, with a washer and dryer that would seem small in most apartments.

Adjacent to the enlisted mess area is the galley. Inside a room about the size of an apartment kitchen, the meals (four per day) are prepared for over 130 officers and men. It's amazing that so much can be done in such a small space. There are all the usual institutional kitchen fixtures (electric mixer, oven, grill, and stewing pots), as well as a pair of refrigerated spaces for food storage. Usually one of these is set up as a deep freeze, the other as a fresh food refrigerator, though for longer patrols fresh food is avoided, and only frozen and dry stores are carried. It is a matter of record that the single most limiting factor to SSN operations is the quantity of food and other consumables. Before a long deployment, virtually every spare nook and cranny is packed with stores—food, soap, paper for the copy machines, dry stores, and, of course, most vital of commodities on board a sub, coffee.

Moving forward on the port side passageway, you encounter the berthing spaces for the enlisted personnel. I should say here that if you have a touch of claustrophobia, this is where it will manifest itself. The three-tall bunks are roughly 6 feet long, 3 feet wide, and 2 feet tall: about the size of a coffin. Each bunk has a comfortable foam rubber mattress with bedding, a light for reading, a blower for fresh air, and a curtain for privacy. All your personal gear goes into lockers on the walls, or the 6-inch-deep trays under the bunks. For the enlisted personnel, this is the total extent of their privacy. This is even further limited, as about 40 percent of the enlisted population has to share, or "hot bunk," their sleeping accommodations. This is because the 688I design just did not have enough room to provide a bunk for each enlisted man. This means that groups of three enlisted men have to share two bunks, with the sleep periods (they sleep in six-hour shifts) rigidly scheduled in advance.

On the starboard side of the boat are the berthing and mess spaces for the senior enlisted personnel, generously known as the "goat locker." Here there is a small seating area about the size of a corner booth at a restaurant, which serves as eating area, office, and conference room for the chief petty officers. Heading aft from here is another aisle of three-high bunks, though these are reserved for each man.

For the officers there is a separate wardroom for eating, studying, and doing paperwork. It is a nicely appointed area with its own pantry for coffee and snacks around the clock. In the middle of the space is a single table that serves as dining table, desk, and conference table. Unlike the commander of almost any other ship in the Navy, the commanding officer does not have a separate pantry to take his meals. He sits with his officers at every meal, giving it the feeling of a family gathering. The submarine service has always been more informal than the surface forces, and this is part of the esprit that makes the "bubbleheads" different from the rest. Commander Jones runs a "loose" wardroom where kidding and friendly ribbing is always welcome. He makes no secret of his love of good seafood, and is a big fan of ice cream. In fact, he is fond of saying that other than having the only private stateroom on the boat, his only command privilege on the *Miami* is choosing the flavor of ice cream for the machine in the galley. He chooses a rather diplomatic French vanilla flavor.

As for the commander's cabin, it is hardly the stuff you might find on the *Queen Elizabeth II*. Located just forward of the enlisted mess, on the second level, it is roughly 10 feet long by 8 feet wide. It is dominated by a combination desk/closet unit in the after portion of the cabin. Against the outside bulkhead is a pair of seats with a small table between them; this unit folds down into the bunk. Commander Jones is proud of saying that it's the best bunk on the boat, and certainly it is the only one that does not have another bunk above and/or below it! On the door to his cabin are three notices. One reads KNOCK AND ENTER and another is, THINK QUIET! IT'S OUR BUSINESS . . . IT COULD BE OUR LIVES. The final one is a copy of Rudyard Kipling's famous poem, "If," not a bad philosophy to advertise if you are in charge of 132 lives and $800 million of the taxpayers' money.

The commander's desk contains a variety of different manuals, a safe for classified documents, and various communications devices to keep him in touch with the rest of the boat. One of the newest pieces of equipment to be added is known as a multifunction display, mounted adjacent to his bunk. This marvelous device, which is tied into the BSY-1 combat system, is a red gas-plasma display showing data on position, course, speed, heading, and depth, as well as modes to show the current tactical situation around the boat. The advantage to Commander Jones is that he can wake for a moment in the middle of the night, reach over and check the boat's status, then roll over and go back to sleep—all without having to ruin his night vision by turning on a light or having to pick up a phone and talk to the OOD. He figures that not having to wake up fully several times is worth several hours' more sleep. And that can be life and death for the boat in a combat situation. A total of eight of these devices are located around the boat in such places as the control room and sonar room.

The Engine—The Reactor/Maneuvering Spaces

If you wander aft from the enlisted mess, past the forward escape trunk and down half a deck, you find the great divide on the *Miami*. This is the entrance to the tunnel aft to the propulsion spaces containing the S6G nuclear reactor (built by General Electric) and the main engineering spaces. It is marked by a number of different warning signs from the DNR, ranging from information on possible radiation hazards, to security notices about just who on the boat is allowed aft of this point. It should be noted that no member of the media, including myself, has ever actually seen an actual nuclear submarine reactor compartment or her engineering spaces. Nevertheless there are a number of things that we do know about these areas, and I will try to share them with you.

The first thing to understand about the nuclear reactor on a submarine is that it has only one real purpose, to generate heat to boil water into saturated steam. Other than that, all of the other parts of a nuclear submarine propulsion system are similar to any other type of steam-powered turbine plant. Its advantage over an oil-fired steam plant is the amount of energy concentrated in the nuclear fuel in the reactor core, as well as the complete lack of any need for air. On a weight and volume basis, nuclear fuel, such as enriched uranium, has several million times the amount of stored heat of a comparable amount of fuel oil. This concentration of energy is what makes all the dangers of handling nuclear fuel worth the trouble. In addition, because of the efficiency of the nuclear "fire," it is possible to build boiler plants that are considerably smaller than comparable oil-fired plants.

The process of nuclear fission is essentially quite simple. Imagine a floor covered with mousetraps. Each mousetrap has, mounted on the striker arm, two Ping-Pong balls. If we imagine a uranium atom as a mousetrap, it is holding on to a pair of attached particles called neutrons much like the Ping-Pong balls. Now if you drop another Ping-Pong ball onto one of the traps and trip it, two balls will fly into the air. This represents what happens when a neutron enters the uranium atom and strikes the nucleus: the atom splits and releases the two neutrons, releasing energy as heat. And when those two fall onto two more traps, these will trip and each throw two more Ping-Pong balls skyward. This will continue to double and double again until all the traps fire off their balls in one final fusillade. This same principle, whereby neutrons strike more and more atoms until all of them finally split, is called an uncontrolled or supercritical fission reaction. And this is what happens when an atomic bomb detonates.

But we don't desire an explosion, we want a slower reaction like a fire in a boiler. Imagine that in our room of mousetraps and Ping-Pong balls, we hang some monkeys from the ceiling. And we train them to grab one out of every two Ping-Pong balls when a trap goes off. This would allow the series of tripping traps to go on for a much longer time. And this is exactly what happens in a nuclear reactor. Instead of monkeys, a reactor uses what are called control rods (made of a neutron-absorbing material like cadmium or hafnium) set to absorb exactly the right amount of neutrons to bring the reaction into controlled or critical fission. This reaction still generates a great deal of heat, which is used to boil water into saturated steam to power the sub's turbines. In this way the same nuclear fuel that can cause a nuclear explosion in an instant can be used to power a ship for a period of years. And because of design procedures that have been tested over a period of decades, the fuel in the reactor cannot explode or even come close to doing so. The DNR takes great pride in the safety record of the boats with U.S.-designed reactor plants, which is perfect.

Most of the heat in the reactor is collected into what is known as the primary coolant loop. This is a series of pipes passing an extremely pure water-based coolant through the core of the reactor. This heat is passed through a heat exchanger to what is called the secondary loop. This is where the water for the steam turbine is actually boiled. Now, the steam created here is not the stuff you get from the tea kettle on your stove. This steam, which is under high pressure, is heated to literally hundreds of degrees and contains a great deal of motive energy. And this is the stuff that turns the turbine blades of the main engines, which feed into the reduction gears, which turn the propeller shaft and the propeller. Quite simple, really!

There are a few small problems with this system, though, and we need to discuss them. The obvious one is the question of how to protect the men aboard from the harmful effects of the reactor's radiation. As we mentioned before, the early Soviet nuclear boats scrimped on shielding and became cancer incubators for the naval hospitals of that now-defunct nation. The answer, in a word, is shielding. The entire structure surrounding the reactor compartment is layered with a variety of different shielding materials.

Between the reactor compartment and the forward part of the boat is a huge tank of diesel fuel, which powers the big Fairbanks-Morse auxiliary engine in the machinery compartment. As it turns out, that fuel is extremely efficient at modulating or absorbing the various subatomic particles that could damage human tissues. In addition, the entire reactor is contained inside a reactor vessel that looks like an oversized cold capsule on end. Surrounding this vessel, as well as inside of it, is a system of layered shielding. While the materials actually used are classified, it is easy to deduce that lead (an excellent gamma ray absorber) and chemically treated plastics (based on fossil fuels) are probably used extensively.

In addition to its extensive shielding, the entire reactor plant has been overengineered. Since its earliest beginnings, the DNR has insisted that naval reactors be built with extremely high safety margins. While DNR will not comment, for example, upon just how much pressure all of the reactor plumbing can take, it is generally acknowledged that the entire reactor plant has been built several hundred percent more robustly than is required (400 percent to 600 percent has been mentioned). In addition, every system has at least one backup and usually an extra manual backup on top of that. The legacy of the *Thresher* loss is this fanatical obsession with safety.

Another area of extreme secrecy is the exact configuration and design of the reactor core itself. In fact, other than the technology used to reduce radiated noise, nothing on the *Miami* is as sensitive as the power plant core. This probably consists of a series of uranium fuel elements formed into plates to allow maximum heat transfer to the primary coolant loop. The fuel elements are probably mounted parallel to each other in a fuel assembly mounted atop a support structure in the base of the reactor vessel. The fuel used is highly enriched Uranium-235, probably 90 percent pure U-235 or better. For those who might wonder, the fuel used in commercial nuclear power generation plants runs about 2 percent to 5 percent pure, and the material used in nuclear weapons is about 98 percent pure. In between each fuel element is room for a control rod (also in the form of a plate and made of a neutron modulator), to control the rate of nuclear fission. Each rod is designed to drop automatically into place between two fuel elements in the event of a reactor problem, thus quenching the nuclear reaction. In addition, a procedure called scram allows the crew or the automated monitoring systems to shut down the reactor immediately, and restart it later if conditions allow.

Around the core circulates the coolant of the primary loop, which feeds the heated coolant into a steam generator. The steam generator directs its steam into a secondary

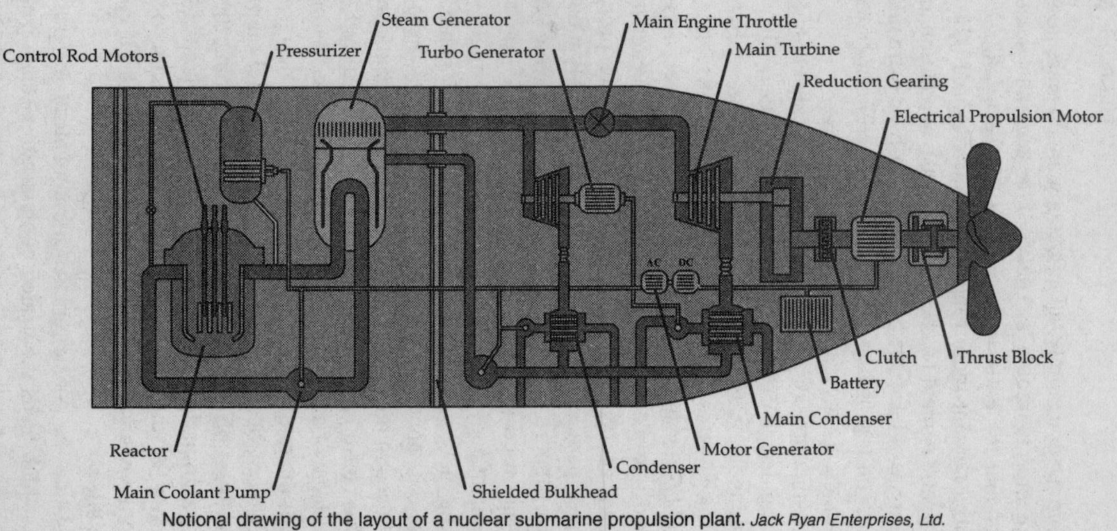

Notional drawing of the layout of a nuclear submarine propulsion plant. *Jack Ryan Enterprises, Ltd.*

cooling loop, which feeds a pair of high-pressure turbines in the machinery spaces, where the steam is condensed back into water and fed back into the steam generator. The turbines feed into a massive set of gears known as reduction gears, which turn the main propeller shaft. In addition, some of the steam is used to turn several smaller turbines that provide electrical power to the boat and its various pieces of machinery.

It may come as a surprise that other than the transit tunnel aft to the main machinery space, the reactor is not manned. The DNR limits the time a man can stay in proximity to the reactor, even how long he might stay in the transit tunnel. The actual control area for the reactor plant and the turbines, called Maneuvering, is located aft in the engine room. While it has never been shown to the press, it probably follows the convention of commercial power plants, with the controls laid out over a block diagram of the reactor/turbine system. This panel is manned at all times, even when the boat is in port and the reactor is shut down (noncritical).

The dominating feature of the machinery space is the deck, or more correctly, the mounting for all of the machinery. While it may seem solid enough, it is in fact a large platform or "raft," which is suspended on mounts on the inside of the hull. The mounts use at least one, probably two, sets of noise isolation mounts. These are like oversized shock absorbers designed to reduce the vibrations of the larger pieces of engine room machinery. The purpose of a raft is to take the noisiest things on the boat and isolate them from the hull, which radiates noise like a speaker into the water.

Mounted on the raft are the two main engines, the boat's electrical turbine generators, and the supporting pumps and equipment associated with moving the boat. Proceeding aft, you see the main propeller shaft leading back to the main packing seals in the stern. In addition there are a number of workbenches, as well as a limited machine shop capable of supporting many small-scale repairs. The size of the main gear, called a bull gear, would preclude repair, but virtually every other contingency in the space could be handled by the engineering team. These crew members, by the way, are recognizable by the different types of radiation monitoring devices they wear. Unlike the film badges worn by those who live and work forward of the reactor, these personnel wear a small dosimeter (which looks like a tiny flashlight), so that any dosage of radiation they receive can be assessed immediately.

To get the power plant started, the engineering officer of the watch orders the personnel at the reactor control panel to retract the control rods to a known position. This allows the core to heat up, causing the coolant to generate steam in the steam generator. From here the turbines are set turning, and so too the reduction gear train. There is a popular notion that the speed of the boat is increased by just retracting the control rods farther from the reactor core. This, in fact, is exactly the converse of what actually happens; the rods are simply retracted to a fixed point and held there. The engineers' main goal is to bring the reactor into equilibrium so that the basic amount of heat going into the primary coolant loop is constant. One can then control the speed of the boat by simply tapping more steam from the steam generator, thereby increasing the steam supply to the turbines. This results in cooling the primary coolant loop more, thus increasing the efficiency of the nuclear reaction, feeding more heat to the steam generator, and increasing the speed of the boat.

Conversely, stemming the flow of steam to the turbines not only slows down the spinning of the turbines, it also takes less heat from the primary coolant loop, and rapidly drops the efficiency of the nuclear reaction, "cooling" it down.

Life Support and Backup Systems

The auxiliary machinery space down on the third level aft of the torpedo room is arguably the most important compartment on *Miami*. Here is located all of the life support equipment, as well as the auxiliary power source. As you enter the space and head down the starboard aisle, you are given a quick introduction to "Clyde," the big auxiliary diesel engine. This is an old favorite of the chiefs onboard, because it is a direct link with the old World War II fleet boats. Built by Fairbanks-Morse, the design dates back to the 1930s and is a scaled-down version of the model used to power all of our submarines during the war. It is reliable and the crew loves it, therefore the name Clyde, as in, ". . . right turn, Clyde!"

While some folks might wonder why such a dinosaur would be on one of the most advanced submarines, remember that not everything always works properly, including nuclear reactors. For example, what would happen if *Miami* was at sea and needed to scram the reactor plant? Restarting a reactor takes a *lot* of power, and while there is a large bank of batteries underneath the torpedo room, it might not prove adequate to completely restart a cold S6G plant. Thus the Fairbanks-Morse engine can provide, through a generator turned by the diesel, enough continuous power to get the tea kettle running again. It has other uses, too. In the event of a reactor casualty, the diesel provides the means for getting home. In that event, the captain orders the engineers aft to lower a small electric outboard motor, which is recessed in the lower hull aft, into the water to provide motive power to get home or to get help.

The diesel engine also has a role in firefighting onboard that might surprise some folks. In the event of a fire, one of the first things the captain might do (assuming this is not in a combat situation) is to surface and start up the diesel. This is because the diesel draws its air from within the boat, and thus it would suck up any air being polluted by the fire. Opening just the fairwater hatches from the control room will completely change the air in the boat in a matter of minutes.

This space is also where the air is made or, more properly, maintained. Several different pieces of equipment in the auxiliary machinery space help to provide the clean, fresh air that can be found onboard. First are the carbon dioxide (CO_2) scrubbers. CO_2 is the gas given off by humans when they breathe and is dangerous when the concentration gets too high. The *Miami* utilizes a chemical scrubber to remove it from the air. The chemical absorbs CO_2 when it is cool and releases it when it is warmed. In addition, CO and H_2 "burners" remove the carbon monoxide and hydrogen gas generated by equipment as well as by cigarette smoking, which is allowed onboard. Finally, filters and dehumidifiers clean the air and help keep it "friendly" not only for the crew but also for the many pieces of equipment—especially electronic—on the *Miami*. In case a fire or some other emergency contaminates the onboard air, a force-fed air supply called the Emergency Air Breathing (EAB) system has attachment points throughout the boat, allowing crewmen with breathing masks to plug in to it and continue their duties.

Other life support equipment includes a device that takes water and electrically "cracks" it into its base elements of hydrogen and oxygen. The oxygen is retained in tanks and released into the boat's atmosphere automatically by the environmental control system, and the hydrogen is vented off the ship from a small port in the aft edge of the fairwater. There is a fresh-water distillation plant that produces something over 10,000 gallons/38,000 liters of fresh water a day. Most of the water is used for drinking, cleaning, cooking, and personal hygiene. Very little water is usually required for the power plant (for charging the cooling loops and steam generators), but the reserve tanks are usually maintained near full "just in case." It should be said that the obsession with water conservation is mostly for contingency purposes. Most COs like to have full tanks of water before they enter a tactical situation, just in case they need to shut down the distillation plant to keep noise down. And from what I hear, some boats just choose to run the distillation plant full-time and let the crew have as much shower time as they want, particularly during runs home. On a normal day aboard *Miami*, the majority of the water produced would go to crew habitability.

Weapons—Torpedoes, Missiles, and Mines

While submarines are useful for covert actions like intelligence gathering and landing special operations forces, it is the threat of what they can do with their weapons that can cause so much fear and respect in an adversary. Ever since Sergeant Ezra Lee tried to sink HMS *Eagle* in Boston harbor back in 1776, just the potential threat of harm from a submarine has been enough to make an enemy stop and consider whether he should move his ships against you.

Today the weapons can hit a wider variety of targets, and they have become even more deadly.

Torpedoes

The torpedo is the traditional weapon of the submarine, and the torpedoes that equip the U.S. SSNs today are truly awesome. For some years now, the U.S. standard torpedo has been the Mark (Mk) 48. This weapon, which first appeared in 1971, has gone through a series of different upgrades, culminating in the Modification (Mod) 4 version, which appeared in 1985. This version, designed as an intermediate upgrade to the next major version, allows for the greater speeds and deeper diving depths of the newer Soviet subs that were appearing at the time. As this book is written, about half the torpedoes being loaded aboard U.S. subs are Mk 48 Mod 4s.

A recent addition is known as the Mk 48 Advanced Capability (ADCAP) torpedo. Manufactured by Hughes, the ADCAP takes the basic Mk 48 package and adds the following new features:

- A bigger fuel tank that provides for a 50 percent increase in range (about 50,000 yards), and a speed of 60+ knots.
- A new data send/receive module, which packs 10 miles of guidance wire into the aft end of the torpedo and 10 more miles into the dispenser in the tube. This allows the submarine to clear the launch point and still guide the weapon.
- A new combination seeker head/computer that uses electronically steered sonar beams to guide the weapon to the target. Earlier versions of the Mk

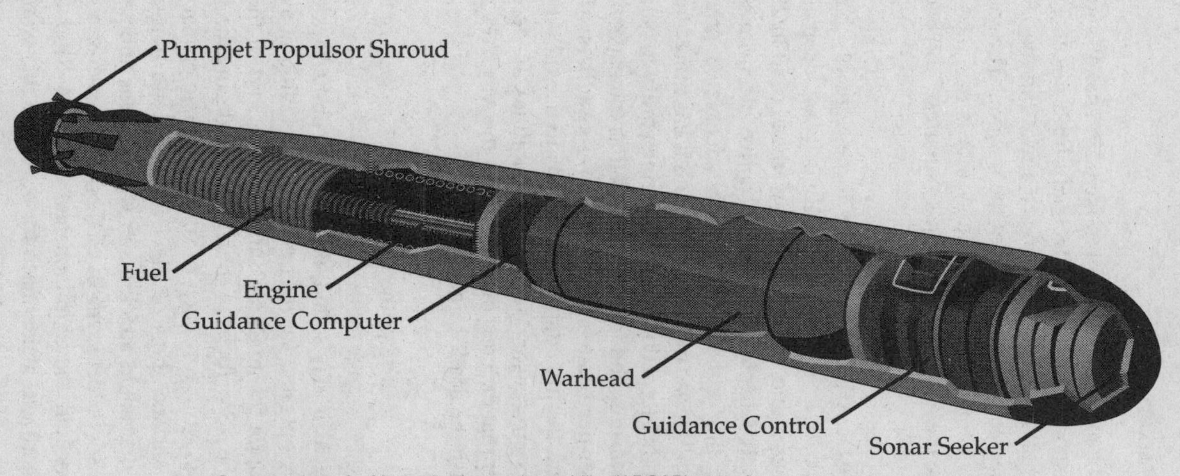

Cutaway view of a Mark 48 advanced capability (ADCAP) torpedo. *Jack Ryan Enterprises, Ltd.*

> 48 (like the Mod 4) used to have to "snake" about their course to search effectively for a target. The head allows the torpedo to see almost all the 180-degree hemisphere ahead of the weapon. The computer controlling the whole system is designed to make the ADCAP the world's "smartest" torpedo.

With ADCAP, the submarine force arguably has the finest torpedo in the world. Not only is it fast, deep diving, and maneuverable, but it has a big warhead (650 lb/295 kg of PBXN-103 explosive) with an active electromagnetic fuse that allows the weapon to be detonated precisely where it will do the most damage. And it has more "brains" than any other torpedo, with an amazing ability to outsmart countermeasures and jamming, as well as the capability to feed seeker-head data back to the BSY-1 system on *Miami*. This allows the fire control technicians to use the ADCAP as an offboard sensor. With such capabilities as these, it's no wonder that the crew of *Miami* calls the ADCAPs in her racks "wish me dead" torpedoes.

Missiles

Strange as it may sound, the nuclear submarines of the U.S. Navy operated for over twenty years without a dedicated weapon for attacking surface ships. Part of the reason was the ASW focus of the SSN force during the 1960s and 1970s. Also, for much of that time their primary targets, the surface ships of the USSR, had no long-range weapons that could attack a sub while it was submerged. But with Soviet deployment of their first sea-based ASW helicopters and the ship-launched SS-N-14 Silex ASW missile, there was a clear need for a weapon that would allow a boat to stand off farther than the ten to fifteen miles a torpedo shot would allow. It had to be launched from a torpedo tube and carried as an all-up or "wooden" round, requiring no maintenance and a minimum of support.

The weapon that was produced was the McDonnell Douglas A/R/UGM-84 Harpoon. This missile, which can be launched by ships, subs, and aircraft, was originally developed to allow patrol aircraft to shoot at Russian cruise missile subs on the surface. First deployed in 1977, it is approximately 17 feet/5.2 meters long, weighs about 1,650 lb/750 kg, and carries a 488-lb/222-kg high-explosive warhead. It utilizes a radar seeker that looks for surface targets and then initiates an attack "endgame" on the target. Packaged inside a buoyant, torpedo-shaped launch capsule, it is fired from one of the normal torpedo tubes and rises to the surface. When it reaches the surface, the nose of the capsule is ejected, and the missile is launched into the air by a small rocket booster. Once airborne, the booster is dropped, an engine inlet cover is ejected, and the small turbojet engine is ignited. The missile then descends to about 100 feet above the surface, and transits to the area of the target ship at a speed of about 550 knots.

The Harpoon can be launched in a variety of modes. These include what is known as Bearing Only Launch (BOL), in which only the bearing to the target is known. There is also a series known as Range and Bearing Launch (RBL) modes, which require both range and bearing. Depending on the range to the target and the amount of neutral shipping in the area, the seeker can be set to RBL-L (Large) for open ocean situations, or RBL-S (Small) for tight, short-range situations. If necessary, several doglegs or waypoints can be programmed into the Harpoon's Midcourse Guidance Unit (MGU), which utilizes a small strapdown inertial guidance system to keep the missile on

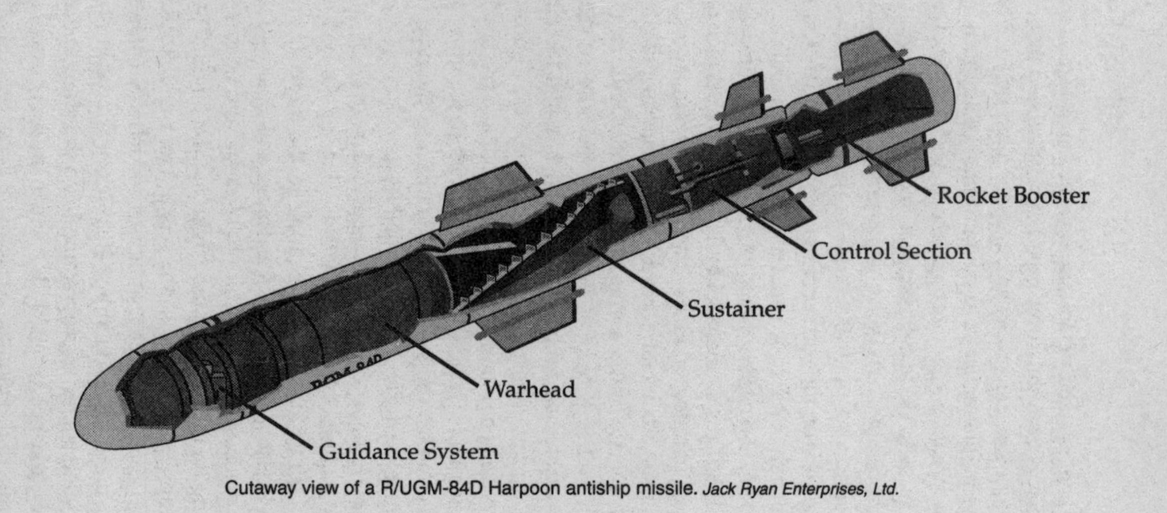

Cutaway view of a R/UGM-84D Harpoon antiship missile. *Jack Ryan Enterprises, Ltd.*

course. For submarines, there is even a self-defense option that allows the defending SSN to shoot the Harpoon "over the shoulder" into a charging surface ship.

Once the missile gets to the target area, the seeker is switched on and begins to search an area shaped much like a piece of pie. If the seeker radar locates a suitable target, the onboard computer does a quick test to make sure it is a valid target (not a wave or a whale), and begins the endgame. The missile descends to an altitude between 5 and 20 feet (depending on the height of the waves) and heads for the target. At the discretion of the *Miami*'s fire control technicians, the missile can be programmed to run straight into the side of the target ship (just a few feet above the waterline), or an optional "pop-up" maneuver can be selected to make the missile plunge deep into the middle of the ship.

In any case, the exploding warhead will tear much of the guts out of any ship up to cruiser size. In addition, any of the jet fuel not used by the missile's turbojet will add to the destruction aboard the target vessel. It is a little-known fact that the warhead of the Exocet missile that sank HMS *Sheffield* in 1982 failed to detonate, but the residual rocket fuel in the missile's motor caused enough of a fire to eventually sink the ship.

The latest version of Harpoon aboard the *Miami* is the UGM-84D, which uses a denser fuel mixture to give it more range (reportedly around 150 NM/250 km). All in all, with some eighteen different countries using it, Harpoon is one of the most successful missile programs ever run by the U.S. Navy.

After the ADCAP, no weapon has done more to make the *Miami* deadly and effective than the UGM-109 Tomahawk cruise missile. Tomahawk is an outgrowth of a loophole that was discovered after the signing of the SALT I arms limitation treaty in 1972. While the exact origin of the cruise missile program is debated, it is generally assumed that Henry Kissinger, then the National Security Advisor, asked the Department of Defense (DoD) to look for classes of nuclear weapons that had not been considered during the SALT I negotiations. After some study, the DoD systems analysts came to the startling conclusion that air-breathing cruise missiles, basically cheap pilotless aircraft with nuclear warheads, would make an excellent weapon to circumvent the terms of the SALT I agreement. They could be launched from ground vehicles, aircraft, ships, and submarines, would be extremely accurate, and would be quite difficult to detect and intercept.

As a result of these studies, a joint project office to develop cruise missile components was started by the U.S. Navy and U.S. Air Force. While both services wound up choosing different models of missile (the Air Force selected a model built by Boeing), most of the components such as engines, warheads, and guidance systems were of a common design. The winner of the Navy competition was the B/UGM-109 model developed by General Dynamics. McDonnell Douglas is the second-source contractor for the missile, called Tomahawk.

The basic nuclear land attack version of Tomahawk, known as B/UGM-109A (also called TLAM-N), is launched into the air by a small rocket booster. Once airborne, a miniature jet engine about the size of a basketball ignites to power the missile at about 500 knots. It flies low to the surface (whether over the open ocean or land), held there by a guidance unit (MGU) being fed by a radar altimeter. The missile is kept on course by the MGU utilizing an inexpensive strapdown inertial guidance system. Once over dry land, the MGU is updated with position data from

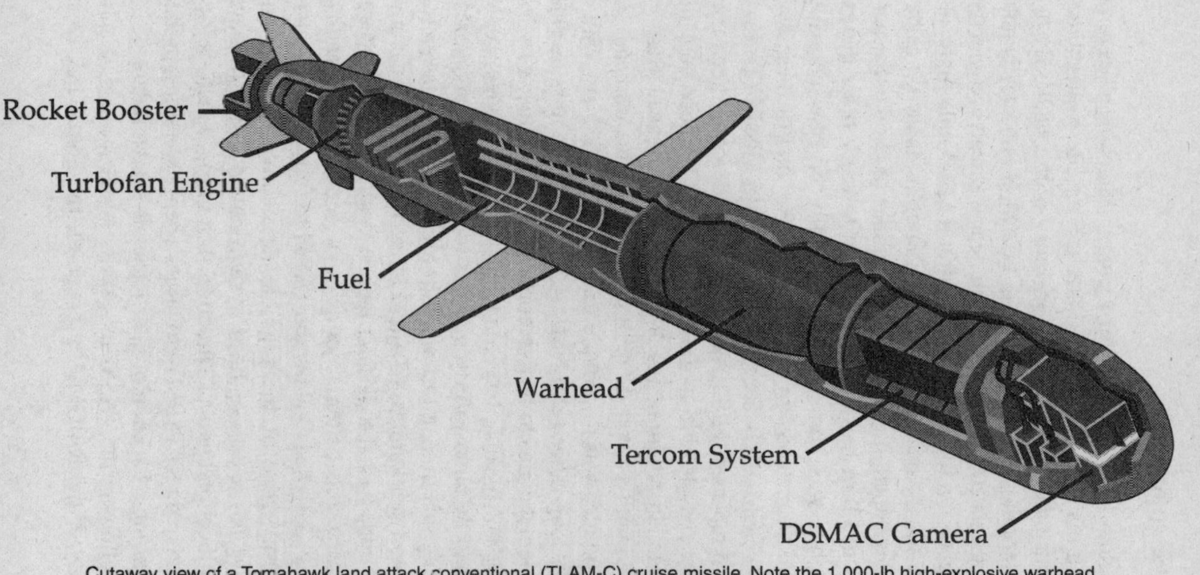

Cutaway view of a Tomahawk land attack conventional (TLAM-C) cruise missile. Note the 1,000-lb high-explosive warhead.

Jack Ryan Enterprises, Ltd.

a system known as Terrain Contour Matching (Tercom), which matches the terrain under the missile with a three-dimensional database in the memory of the MGU. By using periodic Tercom updates, a TLAM-N is normally able to place its 200-kiloton W-80 nuclear warhead between the uprights of a football goalpost after a 1,300-mile flight.

While the nuclear-armed version of Tomahawk was being developed, it occurred to a number of people that perhaps the Tomahawk could be used to carry other things, and thus was born the whole family of conventionally armed Tomahawks in service now. The first of these was the B/UGM-109B Tomahawk Anti-Ship Missile (TASM), which replaced the TLAM-N MGU with a modified radar seeker and MGU from the A/R/UGM-84 Harpoon antiship missile. In addition, the W-80 nuclear warhead was replaced with a 1,000-lb/455-kg high-explosive warhead.

The idea was to provide units of the U.S. Navy with a really long-range (250 NM/410 km) antiship missile. One problem that had to be overcome was the fact that a TASM flying out to hit a target ship at maximum range would have to fly almost thirty minutes to get to the target area. During this time, a fast warship might travel as far as fifteen to twenty miles, so a special series of search patterns was added to the TASM launch and control software. These search patterns comprise a series of "expanding boxes" designed to allow the TASM to fully search the uncertainty zone or the possible target area. In addition, TASM has a passive ESM system called PI/DF (Passive Identification/Passive Direction Finding), which is designed to direct TASM onto larger enemy warships, probably through detection of their large air-search radars.

Following the TASM into service was the largest subfamily of the R/BGM-109 program, the Tomahawk Land

Attack Missile-Conventional (TLAM-C) series. This particular series takes the basic guidance system of the TLAM-N, adds the high-explosive warhead of the TASM, and a new terminal guidance system called Digital Scene Matching (DSMAC). It has a range of roughly 700 NM/1,150 km, and uses the same basic Tercom system to get into the vicinity of the target. DSMAC is an electro-optical system that matches the image from a small television camera in the nose of the TLAM-C to one stored in system memory. This system can even be used at night, with a strobe light on the target during the final approach. Called the B/UGM-109C, it became the first of the Tomahawk series to be used in combat, during Operation Desert Storm.

Several derivatives of the basic TLAM-C include the B/UGM-109D, which replaced the basic high-explosive warhead with a dispenser for 166 BLU-97/B combined effects (fragmentation and blast) submunitions. Called TLAM-D, these Tomahawks are particularly effective against vehicles, personnel, soft targets, and exposed aircraft. A further variant of the TLAM-D, which is armed with antirunway cratering submunitions, is known as the B/UGM-109F. The newest version of Tomahawk, called Block III, incorporates a number of new features such as its own Navstar GPS receiver, a new penetration warhead, an improved engine, and more fuel to bring its range to over 1,000 NM/1,640 km. It should be operational in 1994.

All the various types of Tomahawks can be loaded and fired from any 21-inch/533mm torpedo tube or VLS tube on the *Miami*. Besides twelve missiles in the VLS tubes, additional Tomahawk rounds, as required by a particular mission, can be stored in the torpedo room. This makes Tomahawk the most flexible strike system ever deployed by the U.S. Navy. It also opens a new dimension for the

U.S. SSN force, since now they can join the surface and air forces in striking "over the beach" at significant targets.

The following might be a typical mission load-out for the *Miami*. When preparing to leave for a cruise to the Mediterranean, she might carry a full load of Tomahawks, which would include twelve VLS tubes full of TLAM-C/D variants, with several more in the torpedo room racks. In addition, she would carry a mixed load of Mk 48 Mod 4s and ADCAPs, as well as several Harpoon Block ID anti-ship missiles. There would be no TLAM-Ns, as all of these have been withdrawn from U.S. ships, aircraft, and submarines following President Bush's order in the fall of 1991. Nevertheless, though it is the policy of the U.S. Navy not to deploy nuclear weapons, and they normally refuse to discuss it, the capability does still exist. Also, there would be no TASMs aboard, as the submarine community seems to feel that the Harpoon Block ID is more than adequate to the antishipping task, and the TASMs are hard to get long-range targeting for, on a submarine.

The biggest single bottleneck to effectively utilizing the growing force of TLAM-C/D cruise missiles in the inventory is the preparation of suitable mission plans. Each mission plan has to be developed from a Tercom data base that the Defense Mapping Agency (DMA) has assembled over a period of fifteen years. The data is made into mission plans at one of the Theater Mission Planning Centers (TMPC) located at various places around the world. Here the Tercom data bases are merged with terminal target photos (for the DSMAC cameras), to produce mission plans that can be stored on disk packs on the sub, or downloaded to the sub via a satellite link.

Once the *Miami* has a particular mission plan aboard, the basic plan can be modified on the BSY-1 Command

and Control System (CCS Tac Mk 2) console in the control room. Located adjacent to the BSY-1 fire control consoles, this console can be used to plan and control missions for all the variants of Harpoon and Tomahawk. Should *Miami* not have a plan available in her onboard library, she can use the CCS-2 to develop her own plans. And with the coming deployment of the Block III version of TLAM-C, the requirement for access to a complete Tercom library for mission planning will be reduced.

To launch a Tomahawk or Harpoon, the boat has to slow to about 3 to 5 knots and come to periscope depth. The CCS-2 (or BSY-1 in the case of Harpoon or TASM) console operator powers up and loads a mission plan into a missile loaded in either a torpedo or VLS launch tube. This can be done for as many or as few missiles as the situation requires. Once this is done, the weapons officer inserts a launch key (a holdover from the old TLAM-N days) and presses the firing button. If the weapon is a Tomahawk, it is ejected from the tube (the version fired from torpedo tubes is carried in a tube liner), fires its booster rocket, and away it goes. If it is a Harpoon, the weapon in its buoyant capsule is ejected from the tube and heads for the surface. When it gets there, the booster rocket fires, and it heads for the designated target.

The one problem with all these missiles is that they make the firing submarine extremely vulnerable to detection by aircraft or surface ships, and the amount of noise made by a missile being fired underwater is simply amazing. So it is essential that if the *Miami* is ever tasked with firing a weapon, as the USS *Pittsburgh* (SSN-720) and USS *Louisville* (SSN-724) did during Desert Storm (they fired a total of fourteen TLAM-Cs and TLAM-Ds), she will have to be sure she is clear of any threat during the launch cycle.

Mines

Probably the least appreciated weapons that can be carried by a 688I are mines. These "weapons that wait" are perhaps the most cost effective weapons ever derived for naval warfare. Though most of the mining done by the United States since the end of World War II has been done by aircraft, there may be situations where the stealth and precision of a submarine may be preferred for delivery of these dangerous "eggs."

The first of these is the Mark (Mk) 57 moored mine. It is a derivative of the air-dropped Mk 56 and can be moored in several hundred feet of water. It has a variety of different sensor and triggering systems, including acoustic and magnetic influence fuses. They can be programmed for activation delay or programmed to activate only for certain types and numbers of ships.

Another type is the Mk 67 mobile mine. These are obsolete Mk 37 torpedoes that have been rebuilt into mines that lie on the bottom and wait for a target to drive over them. A submarine might fire them up into a shallow channel, to a distance of 5 to 7 miles. Like the Mk 57, this mine has a variety of different fusing options.

But the crown jewel of the U.S. mine arsenal has to be the Mk 60 Captor mine. This is an encapsulated Mark 46 torpedo, programmed to wait for enemy submarines; when one is detected the torpedo swims clear and attacks the sub. As an added benefit they can be programmed to listen for a certain type of submarine, like a Kilo or Akula. During the Cold War, it was planned to seed Captors along all the transit routes used by the submarines of the Soviet Union. Now they can be used against any of the growing number of countries that have chosen to buy and use diesel submarines in their navies.

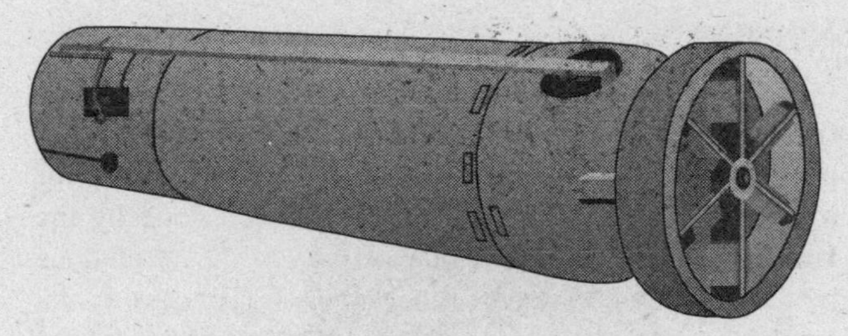

Mark 57 moored mine. *Jack Ryan Enterprises, Ltd.*

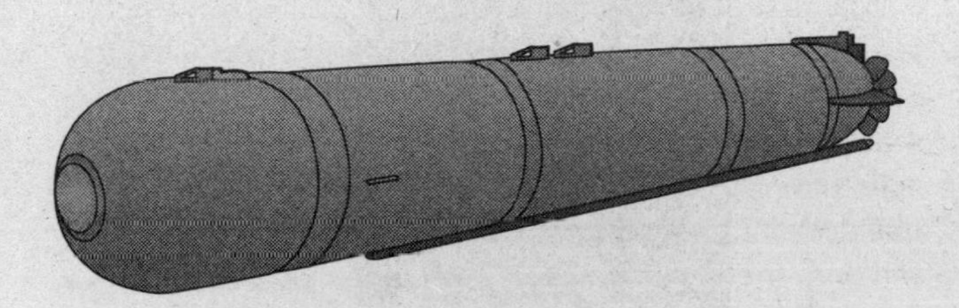

Mark 67 submarine-launched mobile mine (SLMM). This is a converted Mk 37 torpedo designed to be fired from a distance, then to sink to the ocean floor to act as a bottom mine. *Jack Ryan Enterprises, Ltd.*

One of the nice things about mines is that they take up only about half as much space as the other types of weapons a sub might carry. Thus a 688I could carry as many as forty mines and still have a couple of ADCAPs for self-protection. The deployment of the mines is no different from loading and firing a torpedo (the BSY-1 has a mine launch mode), though the position of the mine has to be plotted absolutely accurately, so that it can be swept later. Fortunately the advent of GPS has made this task a bit easier, though efficient use of the SINS system is also required.

All in all, these weapons make for a *very* dangerous quiver of arrows for the submarine force.

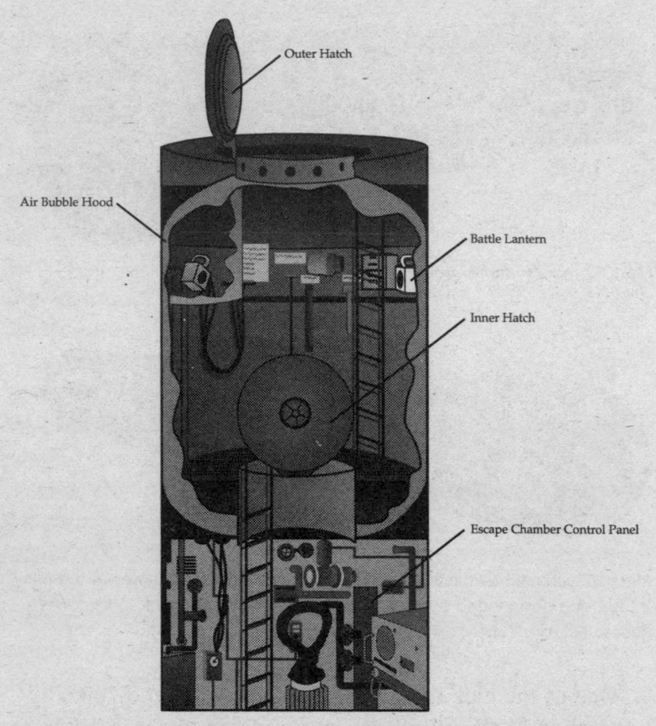

Forward escape trunk of USS *Miami*. Note the air bubble where the crew/swimmers would stand before egressing from the trunk. *Jack Ryan Enterprises, Ltd.*

Escape Trunks/Swimmer Delivery

Wandering aft about 25 feet from the enlisted mess puts you underneath the forward escape trunk. This is a two-man air-lock used for a variety of purposes, though primarily as the main entry point to the forward part of the boat. It is composed of a pressure vessel about 8 feet tall and 5 feet in diameter. At both the top and bottom is a hatch capable of standing as much pressure as the actual hull of the boat. Most often personnel and supplies are loaded through this trunk. There is also another trunk farther back over the aft machinery spaces.

In the event of an emergency the escape trunk comes into its own. If the boat is on the bottom and stable, the normal procedure is to wait for one of the deep-submergence rescue vehicles (DSRV) to be transported to the rescue site. The DSRV then comes down and docks to the collar over one of the escape trunks. It blows out the water from its own docking collar, now held in place by the pressure of the surrounding water. The crew of the DSRV open their own bottom hatch and enter the downed boat through the trunk. Now the crew of the downed sub can come aboard, albeit only about two dozen at a time. This means that if a Los Angeles boat were to go down intact with all her crew alive, it would take something like six trips to get them all off.

If the boat is flooding and the crew must get off immediately, the escape trunk takes on a more vital role, allowing the crew to escape from the boat under their own power. This is done using a Steinke hood, a combination life jacket and breathing apparatus that fits over the head of a sailor. Two at a time the men enter the escape trunk wearing their Steinke hoods. They close the bottom hatch and huddle under an air bubble flange installed in the trunk for such operations. The sailors then charge their Steinke hood air reservoirs from an air port in the side of the trunk, and open a flood valve to fill the trunk with water. While they sit under the air bubble flange, the upper hatch opens. If they are the first ones to escape from the sub, they will have the additional job of pushing a life raft out of the hatch; this floats to the surface and provides some shelter for the men when they get there. Then, one at a time, they duck under the flange and float up through the hatch.

At 400 feet (the maximum depth that the hood can be used), the men will have something like a minute to flood the trunk and get out. Any longer, and they risk getting "the bends" (small bubbles of nitrogen gas that form in the blood) as they rise to the surface. After they have exited, the controller in the area below the trunk closes the hatch via his control panel and begins to drain the trunk for the next pair of escapees. Meanwhile the two sailors literally rocket up to the surface. This would be extremely dangerous (the decreasing water pressure makes them vulnerable to a variety of air embolisms if they hold their breath), but with their heads in the air bubbles of their Steinke hoods, the men are able to breathe normally throughout the ascent to the surface. Once on the surface, they try to inflate the raft and stay together.

One of the other primary uses for the escape trunk, this one far less ominous, is as an airlock for divers and special operations teams. One of the little-known facts about U.S. submarines is that they have, at all times, a small team of divers (usually three to five rated divers) aboard to support the operations of the sub. The diving equipment and other gear is stored in the compartment forward of the torpedo room, near the VLS support equipment room. The divers' jobs include everything from clearing fouled propellers and running gear, to running security checks on the boat before she leaves harbor. In fact, when the *Miami* is in a foreign port she is not allowed to leave the harbor unless she has at least three divers aboard to assist in examining the hull before she gets underway.

The other type of diver-related operation that is conducted through the escape trunk is submerged "lock-out" of special operations teams, such as the U.S. Navy's elite SEAL teams.

These kinds of operations are really not the forté of the 688I and will, until they are retired, be predominately the job of modified Sturgeon-class boats like the *Parche* (SSN-683). Part of the problem is that the Los Angeles–class boats are optimized for speed and are not properly equipped to conduct this kind of mission effectively. Also, the already cramped accommodations of the 688I make it necessary to set up temporary sleeping quarters for the team, perhaps on bunks down in the torpedo room.

In the unusual case of a special operations mission, the boat nears the target of the team and hovers over the seabed. The team then enter the trunk two at a time under the air bubble flange, and follow the same procedure as escaping sailors except with their diving gear. Retrieval is exactly the reverse, with the team reentering the trunk two at a time, closing the hatch, draining the trunk, and exiting through the bottom hatch back into the boat.

The Sounds of Silence—Acoustic Isolation

Silence. That is what has made American boats better than their opponents for over thirty years. It is their armor and their cloak all wrapped up into one vital quality. Nevertheless it comes at a high price and is called a fragile technology—fragile because it is based upon well-understood principles of physics, and because it can be compromised so easily. In terms of military technology, it is one of the crown jewels, in the same category as the ability to build stealth aircraft and nuclear weapons. So effective has this silencing effort been that the latest U.S. SSNs and SSBNs are so quiet, they can effectively disappear in the ocean's background noise.

To make a quiet submarine, the naval architects must take a holistic attitude to the design of the boat and every piece of equipment that goes into it. The key is mounting each piece of equipment that moves or makes noise on something that damps out the vibrations. The transmission of these vibrations—things like the spinning of a pump or the hum of a generator—sends noise out into the hull, where it is radiated into the water. In addition, the rubber decoupling tiles coating the hull help keep noise inside the hull from being transmitted out into the water.

The mounts on the main machinery raft take care of the biggest source of radiated noise. The rest of it is probably taken care of by secondary mounts underneath each piece of equipment (pumps, turbines, etc.), designed to attenuate the specific type of noise generated by that particular piece of equipment. In addition, each piece of machinery is probably designed to be as smooth running and noiseless as America's best mechanical and electrical engineers can make it. For example, the seawater circulation pumps, which are arguably the most noisy devices on the boat, transmit almost no noise in the 688I-class boats. Supporting this is a noise-monitoring system with sensors throughout the boat designed to tell if any piece of equipment or gear is loose or malfunctioning. An added benefit of this system is that it probably is capable of predicting when and how a piece of machinery is going to fail by its acoustic signature (such as the sound of bearings wearing out).

The various techniques used to decrease the radiated noise of American submarines constitute the single most classified aspect of the *Miami* and her sisters. The above description is only the most cursory discussion possible of this incredible technology. In fact, the only real way to describe the magnitude of the achievement is to say that the S6G reactor generates something like 35,000 shaft horsepower[2], yet with all this power the total noise radiated by the *Miami* is probably something less than the energy given off by a 20-watt light bulb. It is for this reason that submariners sometimes refer to their Air Force cousins flying the F-117A stealth fighter as "the junior stealth service."

Life Aboard

So, you ask, what is it like to live aboard a submarine like *Miami*? Well, imagine a combination of living in an oversized motor home and summer camp, and this is a lot of what life in the 33-foot pressure hull is like. Not much room, very little noise, very little news from home, and virtually *no* privacy. Against these "downs" are the esprit de corps of the submarine force, and the knowledge that being a submariner truly makes a man the best of the best in the U.S. Navy.

If you were to go out on a cruise on *Miami*, the very first thing you probably would notice is that you seem to be bumping into *everything* and *everyone* on the boat. This is not unusual for someone new on a sub, and after just a few hours you begin to "think small and thin" so that you can smoothly move around your fellow submariners.

The next thing that comes to your attention will probably be the rather odd working schedule, a watch program that has a crewman working six hours "on" and twelve hours "off." While he is "on," a sailor is standing watch; while "off," he is eating and sleeping, doing maintenance on equipment and systems, and studying for qualification. This creates the unusual standard of a *Miami* "day" being

---

[2] A. D. Baker, *Combat Fleets of the World*, U.S. Naval Institute, 1993, pp. 809–811.

eighteen hours rather than twenty-four. Unfortunately the entire boat takes on this schedule, which tends to lead rapidly to crew members' suffering from sleep deprivation. While in theory a crew member is allowed eight hours of "off" time in a given twenty-four-hour period, this rarely works out into long periods for sleeping. Very quickly one loses all sense of time on the surface and back home, and the sleep that one does get tends to be "on the fly."

As for sleeping itself, this is a relatively comfortable thing to do on *Miami*. With the exception of Commander Jones's stateroom, the bunks for all the officers and men are about the same size, with similar appointments. And while a berthing space is about the same size as a big coffin, once you learn to think small, the space seems quite roomy. With fresh air blowing on your face and a nice foam mattress, falling asleep is really not much of a problem.

What is a problem is the "hot bunking" required for a large portion of the enlisted personnel on the *Miami*. This tends to dominate the schedules of the junior enlisted crew members, with a rigidly set schedule for many of the berthing spaces. If "special" or extra personnel have to be aboard, the crew will lay out extra bunks in the torpedo room over the weapons stowage racks. These are actually quite comfortable, with good headroom, though some folks find the idea of sleeping in a room with literally tons of explosive and fuel rather discomforting. Another problem is the lack of personal stowage space. For those with their own bunks there is a 6-inch-deep stowage pan under each mattress, as well as some locker space. For those having to "hot bunk," three men have to share the space normally allotted to two.

Dining aboard *Miami* is truly a pleasure, as the Navy goes all out to give the men the best chow the taxpayers' money can buy. In fact, because of the limited room for exercise, many of the men actually tend to gain weight on cruise. The food itself is simple but wholesome, with fresh fruit and vegetables becoming the most prized items after a few weeks. The Navy has done some rather clever things to extend the storage life of much of the fresh food aboard. For example, eggs are specially treated with a wax coating to extend their shelf life.

The cooks and their helpers (everyone does an occasional stint of mess duty) work hard to vary the menu and make meals interesting, using a galley about the size of an apartment kitchen. Certainly the culinary highlight of a cruise is the traditional halfway meal of "surf and turf" (steak and crab legs). Unfortunately, by the last few weeks of the cruise every man aboard will be sick of three-bean salad, and dreaming about fresh veggies almost as much as he does about his family.

Those dreams of home and family are always at the center of the submariners' thoughts, though there is very little the Navy can do to give them the kind of communications home that sailors aboard a carrier or frigate might have. The stealth of the modern SSN means that the crew of the *Miami* is almost never allowed to send personal messages home, and news from home is heavily limited and censored. Word from home is limited to a series (about one a week while on patrol) of forty-word messages called "Familygrams." Each Familygram is carefully crafted by a wife, parent, or loved one to give the crewman at sea an idea of what is happening at home. An example of a notional Familygram is seen below:

> 421. DOE LTJG 5/14: REMEMBERED MOTHERS-FLOWERS BEAUTIFUL-THANK YOU. GREAT NEWS. IN CHARGE SUMMER CAMP PROGRAM. THIRTY KIDS. STARTS 24TH. BOUGHT SWIMMING POOL JOHN JR. B ALGEBRA SEMESTER. NO TIME FOR GARDENING, CERAMICS. MONEY FINE-FEW BILLS. SAVING FOR VACATION. MISSING YOU. ILY. JANE

Once the Familygram has been placed into a drop box at the boat's home base of Groton, Connecticut, it is reviewed by personnel at the submarine group for any security problems or personal bad news. Occasionally the message will be returned for an edit or suggested change. As a general rule, no "Dear John" letters or bad news (death, illness, etc.) will be transmitted to the boat.

In addition, when the ship's office on *Miami* receives the Familygram, the personnel will also look over the messages and forward any that look like problems to the captain or executive officer for disposition. The Navy is quite conscious of the sacrifices of those who choose to love and live with submariners, and tries to close ranks whenever there is trouble. As it is, the majority of submariners I have met treasure the Familygrams they have received over the years on cruise. In these notes are news of babies on the way and babies born, birthdays and first words. For the men aboard *Miami* and all U.S. submarines, the Familygram with its "ILY" (I Love You) greeting is the only news they want to hear. It is their sole lifeline to home and "the world."

One of the ways the Navy helps the crew keep their minds off their homes and loved ones is to work them very hard. Every day the officers and men stand watches, maintain equipment, and study. This studying, known as qualifying, takes up almost all the "free" time of a submarine sailor. Since the days of World War II, when the sub force had to expand rapidly, the Navy has always pushed its submariners to gain knowledge and move up in the ranks. There is a ship's library, and video movies on the closed-circuit system, but these tend to be left alone in deference to a sailor's or officer's qualification book. In the enlisted mess, there is frequently a class running in what is known as "the school of the boat." During a visit to *Miami* the chiefs were running an orientation program on the boat's reactor plant—all of this while stores were being packed away and lunch being served.

Another function is the ritual of drills. One of the best ways to keep the skills of the crew honed and their minds sharp is to run daily drills simulating responses to various emergency and combat situations. These may range from fire drills (which are run every day or so) to simulated reactor restarts, to chemical spills ("Otto Fuel spill in the torpedo room" is a favorite!), and tracking drills. The drills are an excellent way to keep the crew from getting bored, and the words "Drill Period" on the boat's plan of the day are both hated and cherished by the crew for the difficult tasks this brings, and the confidence it builds.

The fire drills are quite interesting to watch. Without the facilities and equipment back home at Street Hall in Groton, the chiefs on the *Miami* are hard pressed to simulate the effects of such emergencies. For example, say there is a fire in one of the machinery spaces. The XO and his fire response team move to the compartment where the exercise is being conducted, with all the equipment they would use if the emergency was real. There the fire team find the drill supervision team equipped with gray tablecloths (to simulate smoke), and they must perform to the ship's accepted standards.

Other normal day-to-day functions take on some interesting bents on the *Miami*. Just aft of the drink machines is the ship's laundry, which hardly seems worthy of the title. At about the size of a phone booth, it has a tiny washer and dryer that would hardly be satisfactory in an apartment unit. Here it serves the needs of over 130 officers and men.

Even taking the garbage out has its exotic aspects. Just forward of the enlisted mess on the starboard side is the compartment containing the Trash Disposal Unit (TDU). The compartment contains the TDU (which looks like a small torpedo tube going through the floor), a garbage compactor, a large sheet metal roller, and the supplies necessary to dispose of the garbage produced by 132 men for several months.

How this is done is actually quite fascinating. The first step is to roll a "garbage can" out of pierced sheet metal. This can is placed in the trash compactor and filled with garbage. Usually the *Miami* generates two to three cans a day. When the time comes to dispose of them, each can has a couple of lead weights added to it and is sealed. Then the sonar crew does a complete check of the area to make sure nothing is around that might hear the operation. Because of the noise the cans make as they rattle down the TDU ejector tube, it is normal policy to store full cans if the boat is in a tactical situation requiring extreme stealth. In this case, the cans are stored in one of the refrigerated spaces to keep the smell down. When it's time to eject the cans, the cover to the TDU is opened, and a circular cake of ice is placed inside to protect the ball valve at the bottom. The can is placed on top of the ice, the TDU cover is closed, and the can is ejected much like a torpedo.

The daily life on board the *Miami* is filled with many of the kinds of things that go on anywhere that many men are packed together to do a very tough job. The boat becomes a place of quiet, with words whispered and steps taken lightly. And on those occasions when a difficult mission or operation comes along, the boat continues the same kind of routine, only more so. Anything that makes noise, even routine maintenance, is deferred to keep the noise down.

And how do we reward such devotion? By saying "Well done," and giving them more of the same to do. The life of a submariner is one of a private and personal pride, the kind that comes from being part of an elite club that you cannot buy or beg your way into, and you have to perform "above and beyond" just to stay in.

And then there is the ultimate reward of returning these men to their families and homes. It is said that when a boat is going back to base, the engineers in the machinery spaces have a special setting for "going home." If you have ever seen the incredible spectacle of a warship returning its men to the land, you know why. Every wife and girlfriend has her best on for her man, many with new babies and older children under their arms. If you ever want to know why they do it, look at the loved ones they leave behind in the knowledge that their sacrifices protect those they love most.

America can take pride in the sacrifices of these men and their loved ones over the last forty-five years of SSN operations. Pride for a job well done. Pride in what they are. And pride in what they will do in the future.

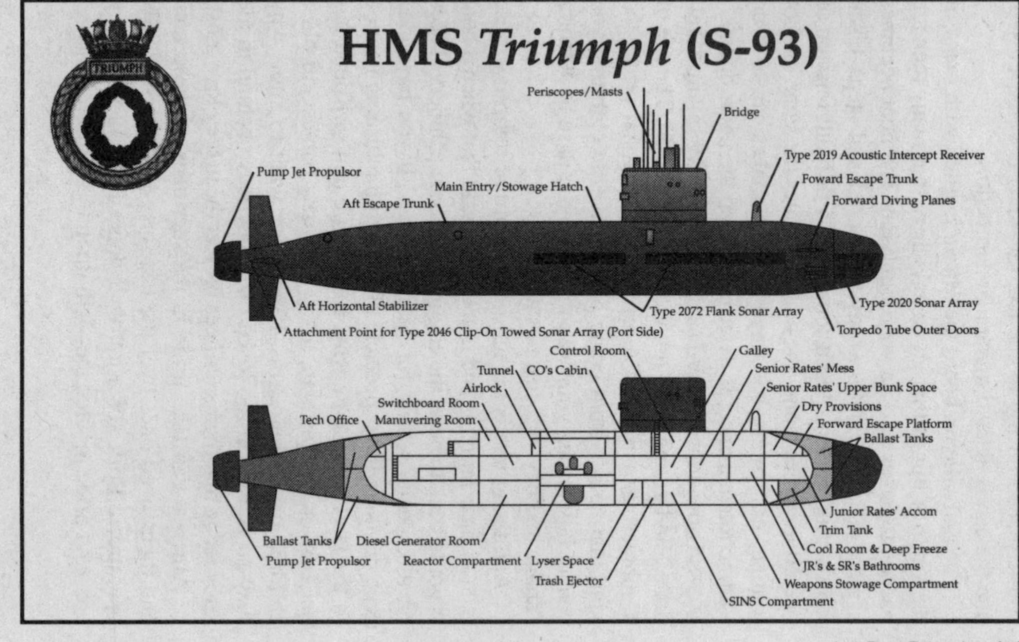

Jack Ryan Enterprises, Ltd.

The British Boats: A Tour of HMS *Triumph* (S-93)

After the United States, the largest builder and operator of nuclear submarines in the western world is the United Kingdom. Currently the British operate a force of twelve SSNs and four SSBNs. In addition, they also operate a small force of four diesel attack submarines. While this may seem like a minor force compared to that operated by the United States, the British fulfill an important role in the structure of NATO. In addition, since they are located much closer to potential points of conflict in Europe and Africa, the responsiveness of their submarine force is multiplied far beyond their small numbers.

If you were to travel throughout the world and talk to submarine captains, and the captains of the surface ships who might have to oppose them, and ask them whose submarines they most fear, you might be surprised. For while everyone deeply respects the Americans with their technologically and numerically superior submarine force, they all quietly fear the British. Note that I use the word *fear*. Not just respect. Not just awe. But real fear at what a

British submarine, with one of their superbly qualified captains at the helm, might be capable of doing.

Royal Navy Submarine History

It is somewhat ironic that the nation that may have the highest quality submarine force in the world has itself been more victimized by submarines than any other in history. It was a British ship, HMS *Eagle*, that was the target of the first attempted attack by a submarine, the *Turtle*. It was also the British who were the intended victims of Robert Fulton's *Nautilus* and John Holland's early submarines built for the Fenian Society. And it was the British who suffered the most during two world wars from the efforts of Germany's U-boat fleets. Certainly no other nation on earth has such understanding of the damage submarines can do.

This is not to say that the British have had an easy time developing their submarine force. The truth is that until the late 1960s, the men who had chosen to serve in the Royal Navy sub force were regarded as pariahs and not considered to be gentlemen by the other line officers of fleet. As far back as 1804, when the British admirals got their first look at Robert Fulton's *Nautilus*, the submarine has been considered a sneaky and "damn un-English" way to fight a war. This opinion had not changed by World War I, though the Royal Navy had begun a modest investment in such craft. Ironically, one of the first customers for John Holland's early submarines was Great Britain, which bought five for experimentation and establishment of her force. Nevertheless, the Royal Navy poured almost all of its funds into a fleet of modern battleships and escorting vessels, keeping funds for submarines scarce. With only a limited force of submarines to use in wartime, the Royal Navy made a point of putting only their most talented officers in command. This wound up paying great dividends, although they did not have the rich numbers and variety of targets that the U-boats had. The exploits of their captains, including the great Sir Max Horton, have become legend in the annals of submarine history, and gave the Royal Navy a tradition they were able to build on.

During the period between the world wars, the British experimented widely in submarine technology. They developed submarines that could carry aircraft and heavy guns, and a variety of new and different power plants. Along with the U.S. Navy, they led in developing the type of submarine that would have the greatest impact in World War II, the long-range fleet submarine. During World War II this force, particularly the "T" class, did the bulk of the damage inflicted by British submarines. In the Mediterranean the "T" boats of the 10th Submarine Flotilla based at Malta sank many supply ships destined for Field Marshal Rommel's Afrika Korps, helping keep him from the oil fields of Arabia. Several of the "T" boats deployed to the Pacific, for which they had been designed originally, to assist in the fight against the Japanese. They even helped in the ASW campaign against the U-boats by sinking seventeen German and Italian submarines.

Another British achievement was in the area of special operations. Throughout World War II, the Royal Navy submarine had an exemplary record, ranging from the insertion of commando teams to the preinvasion surveys of landing beaches. Part of this record includes the use of miniature submarines, called X-Craft, to damage beyond repair the German battleship *Tirpitz* and the Japanese cruiser *Takao*, as well as providing navigation beacons for the British landing forces on D-Day. To this day, special operations are one of the hallmarks of the British submarine tradition.

After the war, the Royal Navy took its share of the German U-boats and technology that it had captured, and began to work on the development of its own "super" submarines. Like the other fleets of the world, the dream of the British submariners was to find a technology that would allow a submarine to travel at high speed, for long periods of time, without having to use a snorkel tube and risk detection. The RN explored the conventional steps of hydrogen peroxide engines and other air-independent systems. Unfortunately they did not invest in the nuclear reactor program that the United States had started in the 1940s, and wound up having to accept that they had bet on the wrong technologies when it became obvious that nuclear power was the future in submarine development.

Because of the special relationship that had been forged between the United States and Great Britain during the war, however, the United States was willing to sell their reactor and power train technology to the British. So in 1963, the first British SSN, HMS *Dreadnought* (S-98), was commissioned into the Royal Navy. She was essentially a Skipjack-class SSN from the reactor aft and a British sub from there forward. And while she made a *lot* of noise, just like her American half-sister, the *Dreadnought* provided the Royal Navy with a foothold into nuclear submarine operations, and the beginnings of a cadre of experienced nuclear sailors. Following the *Dreadnought*, the Royal Navy commissioned five additional SSNs of the Valiant (S-102) class. These new SSNs were contemporaries of the Permit class, and used U.K.-built reactor plants based on the U.S. design.

During this period the British government was trying to find a way of maintaining a credible nuclear deterrent force that would be under *British* control. The force of RAF "V" bombers were quickly losing their ability to penetrate the air defenses of the Soviet Union, and the development of an ICBM force that would reside on British soil was simply beyond the financial resources of Great Britain. So the British government made the decision to buy the Polaris A3 missile system from the United States and build a force of four SSBNs to carry them. Thus was born the "R" class of SSBNs, the first of which, HMS *Resolution* (S-27), was commissioned in 1967. For over a quarter century the "R" boats have provided the United Kingdom with their nuclear deterrent force, helping keep the peace.

By the late 1960s the Royal Navy was beginning to think about expanding further their force of SSNs. Part of the reason was the expanding force of Soviet SSBNs, which had started making themselves known by this time. So a new class of SSN dedicated to ASW tasks was ordered. Called the "S" class, the first unit, HMS *Swiftsure* (S-126), was commissioned in 1973. Contemporaries of the American Sturgeon class, five of the six units built are still in service today.

It was the "S" class boats, along with several of the "V" class SSNs, that provided the Royal Navy with its primary antiship punch during Operation Corporate in the 1982 Falklands War. Three of the boats, HMS *Conqueror*, HMS *Splendid*, and HMS *Spartan*, were the first Royal Navy units to arrive and set up operation in the British-declared Total Exclusion Zone (TEZ) around the islands. They helped give the TEZ credibility long before the surface task force arrived in the area, as well as helping land the first of the special operations teams that were to be so effective during the war. Later, when the Argentine Navy tried to engage the Royal Navy task force, HMS *Conqueror* sank the cruiser *General Belgrano* and scared the rest of their navy back into port, never to come out again.

The year after the Falklands War, the Royal Navy took delivery of what, up to the writing of this book, is the last class of SSNs built, the "T" class. Delivered in 1983, HMS *Trafalgar* (S-107) is the ultimate expression of British SSN design. Still powered by an American-designed reactor (called PWR-1), it was the lead unit of a seven-boat class. And in the area of SSBNs, the Royal Navy has begun trials of their replacement for the "R" class SSBNs, the "V" class. The lead boat of this class, HMS *Vanguard*, will help maintain the British nuclear deterrent force into the twenty-first century. Powered for the first time by a British-designed reactor, the PWR-2, she will carry the same Trident D5 missiles as the Ohio-class SSBNs in the U.S. Navy. A total of four "V" class SSBNs has been ordered.

British Skippers—The Perisher Course

History and tradition are fine, but just what makes a British SSN such a tough proposition to take on? It is, in a word, personnel. As in the United States, the Royal Navy has a submarine school at Portsmouth (called HMS *Dolphin*), which is equipped with a range of classrooms and trainers that would look quite familiar to any U.S. submariner. The British system for manning their submarines, while similar to the U.S. system, has some important differences. It is not all that different in the area of enlisted personnel, though there are some minor differences in the course for enlisted men (women do not serve on Royal Navy submarines as yet). The real difference is for the officers, whose career track is completely different from that of their American counterparts. Starting very early in his career, following graduation from the Royal Navy Academy at Dartmouth, the submarine officer is asked to make a choice of four separate tracks to follow for the rest of his naval career.

One track takes him into the supply branch and can lead to command of a naval depot or a program office. Another is the Marine Engineering Officer (called MEO) track, which allows him to operate a nuclear, steam, or gas turbine power plant. There also is a track for those who desire to specialize in weapons employment. Accepting this option, called the Weapons Engineering Officer (WEO) track, means that an officer can rise to head the weapons department on a submarine or ship. The greatest differences are in the track that leads to command.

For those officers who desire to command one of Her Majesty's submarines, the Seaman Officer's career track must be followed. Much like his U.S. Navy counterpart, the young seaman officer spends his first tour on a submarine qualifying for his "dolphins" and learning how things are done on a submarine. The important difference is that although he spends considerable time watchkeeping and learning the aspects of nuclear engineering that directly concern him, his training is concentrated on making him aware of all aspects of the boat's operations. From the very start of his career, the seaman officer is being groomed for command.

Another difference from his American counterpart is that the young officer spends his entire career assigned to submarines. Shore and "joint" tours are virtually unknown in the British submarine service and are seen as a sign that one may not be suitable for command. As the officer rises through the hierarchy of the wardroom, he becomes first a

Navigator, then a Watch Leader or Officer of the Watch (WL/OOW). During this tour a critical decision about his future is made by his captain and the Chief of Staff, Submarines, at Northwood, England: whether or not to send him to the Perisher.

Perisher is the Royal Navy's submarine command qualification course, which every prospective submarine captain and first lieutenant (the equivalent of the U.S. executive officer) must pass before he can move up into those positions. It is a course unlike anything else in any other service. An American probably would consider it a postgraduate-level course, with an extra helping of stress built in. There is more to Perisher than stress and learning how to drive submarines. It is a test of the trainee's character, designed to tell the Royal Navy whether or not a man is qualified to command one of the most powerful conventional weapons systems in the British arsenal. Probably the closest thing that might be compared to Perisher is the U.S. Navy Fighter Weapons School (Top Gun) at NAS Miramar, California, though Top Gun tests only the skill of a pilot and radar operator, not the ability of an officer to command more than a hundred men. The average Perisher student is in his late twenties or early thirties, with between eight and twelve years of experience in submarines.

About twice a year, ten officers are selected to attend the Perisher course, which is based at the Royal Navy submarine base at Portsmouth. If there are not enough RN officers to fill all ten spots, these vacancies are made available to the prospective captains of other selected navies' submarines. To date, officers from Canada, Australia, Denmark, Holland, Israel, Chile, and many others have taken the Perisher course. The only modification made for these officers is that the parts of the course specifically involving nuclear submarine operations are replaced with instruction on the diesel submarines more commonly found in those navies. Surprisingly, no American officer has ever taken the Perisher course—and it has been run since 1914! I should point out, conversely, that no British officer has ever taken and completed the American PCO course. The two countries have different focuses to their command qualification courses, and both seem satisfied with the products produced.

The faded accounting ledger that is the logbook of every Perisher course since 1922 (the earliest time that they kept records) is filled with a "who's who" of Royal Navy submarine history, including Admiral Sir John Fieldhouse; Admiral Sir Sandy Woodward, who led the RN forces during the Falklands War; and the current senior Perisher "teacher," Commander D. S. H. White, OBE, RN.

Commander White and the other Perisher teachers are the keepers of the institutional memory where command of Royal Navy submarines is concerned. Just two years ago, the Perisher course underwent a significant change in its curriculum, with more emphasis being placed on nuclear submarine operations, long-range weapons employment, and tactics for war at sea. Since that time, the teachers continually try to keep the course and what it teaches as up to date as possible.

The five-month course begins by dividing the ten trainee officers (also called "Perishers") into two groups, each supervised by one of the Perisher teachers. The Perishers visit all the manufacturers of equipment that goes into the RN boats, as well as Vickers Shipbuilding and Engineering, Limited (VSEL) where all the British submarines are currently built. Then they head into the attack simulators to learn approaches to surface targets. After the simulator runs are completed, they head up to the RN Clyde Submarine Base at Faslane, Scotland.

Here the real test of the Perishers begins. Each group of trainees is taken aboard a Royal Navy submarine and begins to do visual approaches on a frigate charging at the submarine. Each trainee gets to do five runs a day for a period of several weeks. As the course progresses more frigates are added, until the Perisher trainee has three of them simultaneously charging at his periscope. The idea is for him to safely operate the submarine, fire off a shot, and not get run over by one or more of the frigates. All the time that a Perisher student is at the conn of the sub, the teacher is evaluating the trainee's reactions and ability to maintain his awareness of the tactical situation.

It is an emotionally brutal regime, with a very high dropout rate. On average, between 20 percent to 30 percent of the Perisher trainees don't make it, and failure rates on individual courses may be as high as 40 percent. Unfortunately, to drop out of Perisher is to never step aboard a British submarine again. When it happens, the teacher's coxswain gives the trainee a bottle of whiskey, and escorts him back to shore.

If the trainee survives the approach phase, he heads into an equally challenging operations phase in which the Perishers play the roles of actual submarine captains on missions. These may include sneaking up on a coastline in the British Isles to deliver a Special Boat Service (SBS) commando team, snap some pictures of a coastline, or practice laying mines. The final phase of the course has the trainees taking part in a war-at-sea exercise, designed to see how each trainee can handle actual command of a boat in combat. When it is all over, and the Perisher has checked off all items on the teacher's checklist to the instructor's satisfaction, he is what every seaman officer dreams of being, a Perisher graduate and qualified to command a Royal Navy submarine.

The Perisher course is a very expensive proposition for the Royal Navy. If it did not already have the assets in place to conduct the course, the cost per individual trainee would be approximately £1.2 million. The human cost is also high. Failed Perishers usually transition into what is known as General Service if they choose to stay in the Navy. If they are lucky, they may even rise to command a frigate or destroyer. But the stigma of being failed Perishers will always follow them.

For all the costs, just what does the Perisher course produce? Arguably he world's finest quality submarine captains. Perisher is the Royal Navy's commitment to making sure that the men who command their submarines are as good as the boats themselves. With only about twenty submarines in the force, they feel they *must* have them commanded by the very best. This is not to say that the U.S. commanding PCO course is not a good course—it is. But by separating the engineering career paths from the service officers at an early point, the future captains can concentrate on being captains, not nuclear engineers. This does not mean that U.S. skippers are not as good as their Royal Navy counterparts, only that the Royal Navy has a procedure in place that automatically selects and qualifies the best of their submariners for their command, not engineering, skills.

Once the Perisher trainee has graduated, he will be assigned as the first lieutenant of a Royal Navy submarine. In the past, when the RN had more diesel submarines, a Perisher graduate could count on getting command of one of these boats directly after completion of the course. Now, of course, all of them do a tour as a first lieutenant. This means that every Royal Navy submarine has *two* men who are fully qualified to command the boat. Once he has done this tour, the officer will likely be given command of his own boat. In, fact, it is not impossible that a good Royal Navy submarine captain might command a diesel boat, an SSN, and an SSBN before he is finished in submarines.

The British like to get their money's worth out of the men they qualify for command, and a really good captain is not done yet. Once a captain has finished with submarines, the Royal Navy frequently sends him to drive ASW frigates such as one of the Type 22 Broadsword class or Type 23 Duke class. By this time a full captain, he is ready to move on to command a task group or naval base, and then, flag rank. This is the *big* difference between the American system and the British. The U.S. Navy system creates superior submarine drivers and engineers; the Royal Navy system is designed to produce pure leaders like a Nelson, Rodney, or Woodward.

THE TRAFALGAR CLASS—A GUIDED TOUR

HMS *Triumph* (S-93) is the seventh and last unit of the Trafalgar class. It is based at the Royal Naval Station at Devonport, near the town of Plymouth in southwest England. She is part of the 2nd Submarine Squadron, which includes the seven "T" boats, and the four diesel boats of the "U" or Upholder class. Ordered in 1986 and laid down at VSEL in 1987, she was launched on February 16, 1991, and commissioned into the Royal Navy on December 10, 1991. At the time this book is being written, the Flag Officer, Submarines, for the Royal Navy is Vice Admiral R. T. Frere, RN. His Chief of Staff, Submarines, is Commodore Roger Lane-Nott, RN. They command the British submarine fleet from the Royal Navy operations center at Northwood, near London.

HMS *Triumph* is the tenth ship (and the second submarine) of the RN to carry the name. Her predecessors carry a total of sixteen battle honors, starting with the battles against the Spanish Armada in 1588. The current *Triumph* is commanded by her commissioning commanding officer, Commander David Michael Vaughan, RN. His first lieutenant is Commander Michael Davis-Marks, RN. Both are Perisher graduates, and each has even commanded one of the cherished "O" class diesel boats before he came to *Triumph.* They are an excellent team, generally considered to be two of the best command-qualified officers in the Royal Navy submarine service. They are aggressive, confident, colorful, and seem fully capable of any tasks that might be asked of them and their boat. Her crew is made up of twelve officers and ninety-seven enlisted men. It is a trim, neat-looking boat with a definite polish to her. Let's take a look for ourselves.

Hull and Fittings

Triumph is somewhat different from the *Miami* in that she is built not so much for speed as stealth. She is smaller than a 688I, at 4,700 tons displacement versus the 8,100 tons for the 6881, and is shorter, around 250 feet/76 meters long. In addition, her hull is more like the classic shape of the *Albacore*, and is somewhat more hydrodynamically stable than the 688I. Her hull is covered in rubber tiles like the 688I, but these are hard and stiff. This coating is anechoic, designed specifically to defeat active sonars that might be trying to get a "ping" off the hull. She may also have a decoupling coating on the inside of the hull to help reduce any machine noise produced internally.

Much like the 688I, her fittings are designed for a minimum of drag, and the only protrusion is the sonar dome for the Type 2019 acoustic intercept receiver forward of the conning tower. Her dive planes are recessed in the forward part of the hull, and she has a fairly conventional set of cruciform tail surfaces aft. At the tip of the vertical stabilizer, the Type 2046 towed sonar array is attached. Unlike the 688I array, this unit is clipped on, not rolled out. This means that it has to be attached and removed whenever the sub enters or leaves port. The 2046 is roughly analogous in capability to the American TB-16.

Though it is not obvious when she is sitting at dock, the most noticeable difference from the 688I is that the boat has no propeller. Instead, *Triumph* is equipped with a device called a pumpjet propulsor. If you could see her in drydock, you would see what looks like a lampshade attached to her stern; this is the pumpjet. This device works like a ducted fan, to push water aft and drive the boat forward. The advantage of this system is that it is somewhat quieter than a propeller, and it operates more smoothly. By the way of example, *Triumph* can speed up from 5 to 18 knots without its crew feeling *any* vibration from the shift in speed. So efficient is this system that the U.S. Navy is planning to use pumpjets on all their future SSNs, including the Seawolf class.

Conning Tower

The conning tower of *Triumph* is much like that of the *Miami*, except that hers has somewhat more room. In fact, there are two separate positions for lookouts and officers to work topside. There is the usual array of periscopes and masts, including a huge dome for the Racal UAP ESM system. Both of the periscopes appear to be RAM coated to keep down their radar signature. Getting down the conning tower trunk into the control room is, if possible, tighter even than on *Miami*. In fact, almost everything on *Triumph* seems to be about three-fifths size compared to *Miami*—sort of like the difference between Disneyland in California and Walt Disney World in Florida!

Sonar Room

If you drop down the ladder into the control room and take a U-turn to the left, you will be in the sonar room of the *Triumph*, where all the equipment and displays for the sonar systems are contained. I should say here that the British have nothing like the BSY-1 combat system in service right now. There is a plan for a system called the 2076 in a few years, but right now, all contact data handed off between sonar systems is done manually. The sonar suite on *Triumph* might be compared favorably to that on a Flight I Los Angeles-class boat. The various sonar systems include:

- Type 2020, the main sonar array (both active and passive) in the bow of the boat. Unlike the dome sonar on *Miami*, it is composed of an array of elements around the "chin" (conformal array) of the boat. It can track several targets at once, and can pass data directly to the fire control system. One of the more interesting features is the "captain's key," which must be inserted in a slot in the 2020 control console before the active mode can be used. It is equipped with a special signal processor, Type 2027, which (if the tactical situation is right) can automatically calculate ranges to the target and feed the data to the fire control system.
- Type 2072, the new flank array (passive listening only), which can only be described as *huge*. It is designed to detect broadband targets at long range.

- Type 2046, the "clip-on" towed sonar array (passive listening only), attached to a tow point on the tip of the horizontal stabilizer. It is capable of detecting both broadband and narrowband signals.
- Type 2019, the acoustic intercept receiver for detecting active sonars and torpedoes. This is a French system that is manned, as opposed to the automatic operating mode of the U.S. WLR-9.

The sonar systems on *Triumph* provide excellent coverage in both spectrum and azimuth. Only the lack of a fully integrated combat system and the TB-23 towed array system keeps it from being the equal technically of the BSY-1.

Control Room/Fire Control/Navigation

If you duck back around the corner where you came from originally, you may be surprised to find that the landing for the conning tower ladder has now been converted into a chair for Commander Vaughan. From this position, he can view the repeater for the sonar systems, the fire control consoles in the track alley, and the plotting area. Just aft are the two periscopes and the mast for the UAP ESM system. The scopes are first-rate, with the CK 034 search scope easily being the equal of the American Type 18. It is equipped with readouts for the ESM receiver mounted on top of the mast as well as a 35mm camera for taking photographs. The CH 084 attack scope, which has a very small head (to make it hard to detect), is also equipped with a low-light TV camera. Both are very quiet when raised, and have excellent optics. Two differences are the use of a split image rangefinder, as well as more automated controls.

The fire control alley is equipped with six positions for fire control technicians. The system is set up to track and engage several targets simultaneously. The screens are round, red- or amber-colored plasma displays; a light pen is used to designate the targets and move between the various operating modes. All the fire control solutions are generated automatically, and there is no manual TMA solution being plotted to back up the automated system. The British seem to prefer this because they believe that most engagements will probably be at relatively short range. This is like what they might encounter with a diesel boat, in which the reaction time for getting the first weapon in the water is the deciding factor. Thus the sonar/fire control fit of the *Triumph*, as well as the training of the crew (and especially the captain in his Perisher course), is a reflection of the current RN combat doctrine.

Traveling aft from the track alley, you come upon the two plotting tables, called SNAPS tables. These are automated and can be fed with plotting information from the fire control system and navigational aids. In addition, they can make use of standard navigational charts, the coordinates of which are stored in the computer's memory. Supporting the navigator is a Navstar GPS receiver, as well as a SINS system (the gyro compartment is down in the third level portside) to help keep *Triumph* on course.

Across the control room to the port side, you find the ship control area. It is laid out similarly to the one on *Miami*, the main difference being that the British have automated the control system so only one man controls both the bow and stern diving planes from a single position. The ballast control panel is to the right of the ship handling position, with the diving officer seated behind them. The boat dives in about the same time as the *Miami*, though she seems to be somewhat easier to trim. *Triumph* handles extremely well, able to turn at over 1 degree per second with only a moderate rudder on. She also speeds up and slows down very quickly and smoothly, with no noticeable sound or vibration as she changes speed. It is the pumpjet that makes most of the difference in noise and vibration over a propeller system like that on the *Miami*. Also, her hull shape is somewhat better from a maneuvering point of view.

The ESM/Radio Spaces

Aft of the plotting area is the radio room. The British communications capabilities appear to be quite similar to those of the *Miami*, though it appears this system may not have an ELF capability. Just aft of the ship control is a door marked RADAR WARNING ROOM. This is the space where the readouts for the ESM system and communication intelligence (Comint) systems are located. Both systems are fed out of the mast antennas, especially the big ESM dome. These are really impressive systems, and are clearly a great deal more capable than a standard 688I. This is not to say the U.S. Navy and the Royal Navy do not have boats specially configured for ESM/Comint purposes; they do. But if I were an American admiral planning to use a sub to monitor radio or radar activity off a hostile coast, and I did not have one of those special boats, I might just ask the British to borrow a Trafalgar-class boat for the mission.

The Engine—The Reactor/Maneuvering Spaces

Aft from the control room, you walk under the main access hatch to the deck, and into the access hatch for the reactor space. As with the *Miami*, visitors are not allowed to enter this space. The *Triumph*'s reactor, called PWR-1

(Pressurized Water Reactor-1), is derived from the American S5W plant. Therefore the British have to abide by all of the procedures and security regulations set down in a 1958 joint RN/U.S. Navy agreement. The PWR-1 supplies about 15,000 horsepower, translating into a top speed of about 30 knots when she is at depth. As far as layout, the machinery spaces are roughly equivalent to those on *Miami*, with two of everything (turbines, motor generators, etc.) except for the main power train.

Living Spaces

Coming back forward on the starboard side is the captain's cabin. The accommodations for the commander of a British SSN are positively Spartan by U.S. standards, with the cabin being only about a third the size of that on the *Miami*. On the forward end of the cabin is a small desk, with a single bunk along the outer bulkhead aft. Maximum use is made of the space, with a bookcase built over the end of the bunk.

Commander Vaughan likes to add a few homey touches to his cabin, like a pile of books on naval warfare (how pleasing to find a hardcover of *The Hunt for Red October* on top!) in the bookshelf, a small sound and video system paneled into the bulkhead, and a Thomas the Tank Engine bedspread, courtesy of his son. While it is somewhat cramped, and he does not even have a head to share with Lieutenant Commander Davis-Marks, he likes it. It is close to the control room, and he can get to his action station in just a matter of seconds.

If you proceed down the accommodation ladder to the second deck, you find the rest of the living spaces. Over on the port side are the officers' quarters and wardroom. The first lieutenant and the navigator share the single two-man cabin, with the rest of the officers sharing spaces with three-high bunks. There is a single lavatory for the officers in the passageway leading to the officers' wardroom. There are the usual amenities of a stereo and video system, as well as plenty of storage for the liquid refreshments that make the Royal Navy seem so much more civilized at times than the U.S. Navy. A small pantry serves the officers' wardroom, though all the food is cooked in a central galley serving all the men on the boat.

The rest of the crew eats and assembles in a pair of small mess areas (senior and junior ratings) on the starboard side of the second level. They are just as comfortable as the officers' wardroom; the senior rating mess has the added luxury of a bar with both Foster's Lager and John Courage on tap. Like the officers' wardroom, both are equipped with stereo and video systems.

The berthing areas are split (senior and junior ratings), with access for all of them located on the second level. Again, they are three-high bunks with stowage trays for personnel gear. As on the *Miami*, there are more junior enlisted personnel than bunks, so some "hot bunking" is required to fit everyone in.

Life Support Systems—The Machinery Spaces

Unlike the 688I-class boats, in which it is all located in one compartment, the Trafalgars have their life support equipment scattered in a series of different spaces in various parts of the boat. The CO_2 scrubbers and the oxygen production plant are down in a compartment on the third level forward, surrounded by an acoustic enclosure. Up on the second level, just above the scrubber compartment, is the air-conditioning plant, also in an acoustic enclosure. Up on the first deck forward, in the same compartment as the forward escape trunk, are the CO/H_2 burners that are used in the event of an emergency. The main H_2 burners are located on the second deck. The two auxiliary diesel engines are located aft in the engine room. The reason for spreading these different pieces of equipment out around the boat is to put them in places where they can be most effectively isolated, from a noise standpoint.

Weapons—Torpedoes and Missiles

Down on the third level and forward, you come to the torpedo room, which the crew calls a "bomb shop." Here are stored the various weapons that arm HMS *Triumph.* She is equipped with five 21-inch/533mm torpedo tubes (two per side, with one going out under the chin of the bow) and can store twenty-five weapons in the compartment. The torpedo tubes utilize a water ram system similar to the one on *Miami*, and use a similar loading system. The fifth tube makes it possible to fire a salvo of four weapons of one type, for instance, while still having one weapon of another type in reserve.

Currently the RN is deploying two different types of torpedoes. One is the Mk 24 Tigerfish Mod 2, which is an electrically powered wire-guided torpedo designed primarily for ASW work. It has a 200-lb/91-kg warhead, a maximum speed of 35 knots, and a range of 22,000 meters at 24 knots. It is very quiet (the British captains are fond of calling Tigerfish the stealth torpedo), though thc small warhead makes it less effective for shooting at surface vessels.

Replacing the Tigerfish is the new Spearfish torpedo, which has a much larger warhead (660 1b/300 kg), comparable range (approximately 13 miles/21 km), and a maximum speed of around 60 knots. This torpedo is a monster, with many of the same kinds of guidance improvements and capabilities as the Mk 48 ADCAP.

In addition to the torpedoes, the RN deploys a version of the UGM-84 Harpoon antiship missile to give the *Triumph* a long-range antiship capability. Called Royal Navy Sub Harpoon (RNSH), it is equivalent to the U.S. Block 1C version of the missile.

While the *Triumph* does not deploy quite the variety of weapons that *Miami* does, one should remember that the British boats do not pursue the same role and missions as the U.S. fleet. And while the Royal Navy captains might like a weapon equivalent to the Block ID or Tomahawk cruise missiles, budget constraints will probably force them to be satisfied with what they currently have. Nevertheless, they are already capably armed and quite deadly.

Escape Trunks/Swimmer Delivery

Much like the *Miami*, the *Triumph* is equipped with a pair of escape trunks for emergency transfer to a DSRV, swimmer delivery, or emergency ascent escape. There is a two-man escape chamber in the forward machinery space on the first level, as well as aft in the machinery space. These chambers are designed to allow emergency escape from depths down to 600 feet/183 meters when used in conjunction with the RN Mk 8 egress/exposure suit. This suit, which uses the same kind of air reservoir breathing system as the American Steinke hood, provides the user with an insulated suit for survival on the surface. So effective is this system that test subjects have been able to survive for up to twenty-four hours in water simulating North Atlantic conditions. Although the British operate in areas where the water is, on average, shallower than that where the U.S. subs operate, they still train all their submarine personnel for deep-water egress. This is regularly practiced in a tower at their submarine school in Portsmouth.

Acoustic Isolation

The Trafalgar-class submarines, much like their American 688I counterparts, are designed to be extremely quiet. And while the British seem to be using many of the same quieting techniques and equipment, there do appear to be a few interesting features. Like *Miami*, the *Triumph* appears to use a large machinery raft with isolation mounts for all the large pieces of equipment (turbines, generators, etc.). Even the shaft that is connected to the pumpjet propulsor has a flexible mounting to keep down bearing noise.

As we discussed earlier, many of the noisier pieces of equipment seem to be set in their own acoustic enclosures. In addition, all the electronic equipment is set on leaf spring mounts to provide protection against the shock of a nearby explosion, as well as some sound isolation. *Triumph* also has a fairly extensive self-monitoring noise system, both to detect any untoward noise as well as to help locate any pending failures. *Triumph* is also equipped with systems to reduce the risk of detection from the boat's magnetic signature, as well as reducing the electrical field generated by the corrosion of the boat in seawater. All in all, the *Triumph* is probably the equal of the *Miami* in noise reduction.

Damage Control

A hallmark of the British character is their power of understatement and reserve. Yet if there is one thing that personnel on board the *Triumph* are fanatical about, it is damage control, particularly firefighting. The British experience with fire during the Falklands War in 1982, specifically the loss of HMS *Sheffield* and RFA *Atlantic Conveyer* to uncontrolled fires, has left a permanent impression. This shows in the design of their boats, which have the ability to isolate compartments and flood them with Halon. Virtually every electronic equipment rack has a port to inject CO_2 gas to snuff any electrical fire. Like the 688I, *Triumph* has an EAB system with forced-feed air masks for every man on the crew. And then there are the firefighting tools themselves.

Their firefighting crew suits are made of chemically treated wool, which they say provides better insulation against the heat of a compartment fire, with protection as good as that of Nomex. Instead of the EAB masks or an OBA to breathe, the RN uses a compressed-air cylinder pack (called a Scott Pack) to provide breathable air to their firefighters. They are equipped with the same kind of thermal imager as the U.S. Navy has, as well as infrared fire detectors (which look like flashlights), and a full array of fire extinguishers, air test kits, and first aid kits.

The crown jewel of the *Triumph*'s firefighting capability is their fixed AFFF (Aqueous Fire Fighting Foam) system. One of these is located on every level of the boat forward of the reactor, and I assume they are also back in the machinery spaces as well. This system, which looks like a small water heater, mixes seawater with the AFFF mixture and feeds it through a pressure hose. Crew members on *Triumph* indicated that they could lay down over 100 gallons/377 liters of AFFF slurry per minute with this system, which compares well with the still very effective AFFF fire extinguishers used on the *Miami*.

Life Aboard

Life aboard *Triumph* is not all that different from on the *Miami*. Though the food is a little different (cheese buns for lunch and curry salad dressing are normal), the diet is healthy and hearty. The cultural difference between the two services appears in the attitude toward alcohol. Unlike the U.S. Navy, the Royal Navy still allows their crews to have beer and wine aboard (the daily "tot" of Pussers Rum is unfortunately no longer served to the ratings). The attitude of the Royal Navy leadership for over six centuries has been that if a man is responsible enough to go to sea with its risks of quick death and isolation, then he should not be deprived of the basic pleasure of a drink if he should want it. In reality, most of what is carried aboard is consumed while in port; most men just don't drink at sea while they are working.

Other aspects of the *Triumph* lifestyle closely parallel that on *Miami*. Water is in short supply, and Navy showers are the rule. The crew uses many kinds of equipment, like the TDU, which any American submariner would feel quite at home with. Watches are roughly the same, with the same problems of having to "hot bunk." The daily routine includes lots of drills of all varieties, ranging from damage control to tactical drills. As for messages from home, the RN seems to follow the U.S. practice of "Familygrams," though probably not quite as often. It is a good life at sea, and the men enjoy it.

Roles and Missions

The folks in the U.S. Navy Undersea Warfare Office (N-87) call them "Roles and Missions." Whatever you call them, these are the tasks that are currently defined for nuclear submarines. Up until very recently, though, just discussing them was cause for extreme discomfort (based upon security regulations) on the part of the senior leadership of the handful of navies that operate SSNs. Now, because of the Cold War's coming to an end and the need to justify the costs of building and operating submarines, those same leaders are letting the world have a peek at just what their boats have done, and still do. In some cases, they are acknowledging for the first time missions that have been conducted for decades. Let's take a look.

Mission #1—Antisubmarine Warfare

The premier ASW platform is and probably will remain another submarine. The reasons for this are defined by the basic advantage of the submarine over other antisubmarine platforms. Environmental factors define the sub's ability to hide. Water temperature, the location of thermocline layers, variations in salinity, and ambient noise sources all are part of the three-dimensional realm of the submarine. The sub lives in that environment and monitors it constantly. Surface ships and aircraft can use their instruments to take snapshots, but they cannot have the broad view that a submarine commander has. Just as ground-based surface-to-air missiles and antiaircraft guns can impede but not deny aircraft the use of the sky, so can surface warships not control the depths of the sea. That's the job of the SSN.

Tactical Example—Stalking a Russian SSBN

They're still out there. They're called boomers in the U.S. Navy, bombers in the Royal Navy. They are the fleet ballistic missile submarines, really creatures of the past Cold War era, but they still sail, and their missiles must be aimed at something—what that something might be, their owners do not say. The Russian ones are probably aimed at the United States, and the American ones at Russia, rather in the manner of a "default" setting on a computer or washing machine. One Russian boomer captain was recently quoted as saying that the target packages on his boat's missiles had not changed, and in fact they might be aimed at some of the nations currently supplying aid to the CIS (Commonwealth of Independent States). Until such time as these dinosaurs are relegated to the past, it is only prudent to keep an eye on them, and that is one mission of the SSN. When a Russian/CIS (formerly Soviet) SSBN departs its home port on the Kola Peninsula, waiting out at sea (possibly in a depression in the sea floor called a "tongue of the ocean") will be a NATO SSN. Probably. Almost certainly, in fact. The mission of the SSN and her crew will be to shadow the Russian SSBN.

The mission is not exactly a friendly one. Should a sudden crisis arise, the SSN's job is to close and destroy the missile boat before she can launch her birds. Short of that exigency, the SSN remains in trail, listening. There is much to learn. Probably the SSN's CO knows the name (or hull number) of the boat he's watching, and he observes the other CO's habits to add to what we already know. He'll listen to the boat, determining her unique mechanical characteristics so that other SSNs can identify her by her acoustic signature. Other observations will tell us much of the quality of the crew, changes in Russian operational doctrine, and from the boat's day-to-day routine, drills and readiness.

It's not quite that easy, of course. Soviet SSBNs are frequently accompanied by their own SSN guardians. Thus the Western submarine must track—and evade detection by—two adversaries who themselves have carefully thought-out routines for dealing with a potential shadower. This can be as simple as running the boomer at high speed toward her protecting SSN, forcing the trailing boat to move quickly herself and so make more noise than the U.S. skipper might wish. Noise is death in this business, and as important as the mechanical characteristics of the platform are, the commander with the most brains has the ultimate advantage.

The mission may be something from the past, but its immediacy hasn't changed. The warheads on those missile submarines are still real. Their aiming points are unknown, but so long as they exist, and so long as men can change their minds, they represent a danger to America and her allies. The smart move is to eliminate the warheads through diplomatic means. Until that happens, eliminating them in other ways will continue to be an option that our leaders will wish to have at their disposal.

So just how does one hunt such a beast? First you must learn its habits and characteristics, and like everything else in this world, the characteristics of the Russian boomer fleet are rapidly changing. With the drawdown in the CIS fleet, and the stipulations of the new START-II arms control treaty, the force of Russian boomers is becoming smaller. By the turn of the century they will probably have only fifteen to twenty missile boats altogether. The ones they keep are going to be the newest, most quiet boats in their fleet. This means that a Western SSN commander is likely to be hunting either a Delta IV or Typhoon-class boat. Both these types of submarine have the latest in quieting technology available to the CIS Navy. To the SSN commander hunting one, this means that even with his advantage in acoustic detection and tracking, which used to allow him to detect and track a target at ranges of tens of thousands of yards, now it's likely that solid contacts will be obtained at ranges of thousands of yards.

Another problem for potential hunters of Russian SSBNs results from the manner in which they are employed and deployed. One of the early goals of missile designers in the former Soviet Union was to make the ranges of their sub-launched missiles as long as possible. It is an acknowledged fact that CIS boomers can launch their missiles at targets in the continental United States from alongside piers at their Kola Peninsula bases. Consequently the only reason the Russian leadership has for moving them is to hide them against possible attack by aircraft or missiles. And like prized jewels, the CIS Navy tends to place them in the maritime equivalent of bank vaults: the "boomer bastions."

Bastions were originally created to place Soviet SSBNs beyond the reach of Western ASW forces. While the actual location and layout of a boomer bastion is a highly sensitive subject in both the Pentagon and the Kremlin, the basic concept is quite simple: an SSBN is placed in a patrol area that is highly defendable and as remote from Western operating areas as possible. The Barents Sea, the Kara Gulf, the Sea of Okhotsk, and even sites under the polar ice pack have been suggested as possible bastion areas. This may mean the SSBN is placed in an area with entrances that are easily defended, or it might be surrounded by a belt of ASW mines. In addition, it probably is aggressively defended by Russian attack submarines, maritime patrol aircraft, and, if available, surface ASW groups.

Clearly, a boomer bastion is not the kind of target a carrier battle group is going to take on. In fact, a modern SSN is the only platform that can even begin to think about penetrating the bastions and pursuing the Russian SSBNs contained therein. Back in the early 1980s the U.S. maritime strategy had NATO trying to actively pursue the Soviet boomers in their lairs. Today the task is made more difficult by the decreased size of the NATO SSN force and the greater stealth of the CIS SSBNs.

Let's assume that Western intelligence services manage to find a boomer bastion. The method is not particularly important—it might be a satellite photo of a missile boat breaking through the polar ice during a missile drill, or radio traffic from a supporting surface group. For our purposes, though, we will assume that the target is a Typhoon-class SSBN being protected by an Akula-class SSN. Their bastion area is a parcel of the Barents Sea that overlaps the polar ice pack in what is called the marginal ice zone. The interface between the polar pack and the marginal ice zone is an extremely complex acoustic envi-

A Los Angeles–class sub breaks the surface during an emergency blow drill. *(Electric Boat Div., General Dynamics Corp.)*

German sub U-58 alongside the USS *Fanning,* having her crew removed after being forced to the surface in November 1917. *(Official United State Navy Photo)*

U-158 sinking by the stern after being bombed by U.S. carrier planes in the Central Atlantic in August 1943. *(National Archives)*

Crewmembers of the USS *Skate* on deck during Arctic operations in March 1959. *(Official United States Navy Photo by Lieutenant Meader)*

Strategic missile sub USS *Ohio*, SSBN-726, underway during sea trials. *(Official United States Navy Photo by PH2 William Garlinghouse)*

A Los Angeles–class submarine on the surface. *(Electric Boat Div., General Dynamics Corp.)*

The business end of an Mk 48 ADCAP torpedo. The black cover is the acoustic "window" of the torpedo's seeker head. *(John D. Gresham)*

An Mk 48 ADCAP torpedo on a loading tray is pushed forward into the #4 torpedo tube. *(John D. Gresham)*

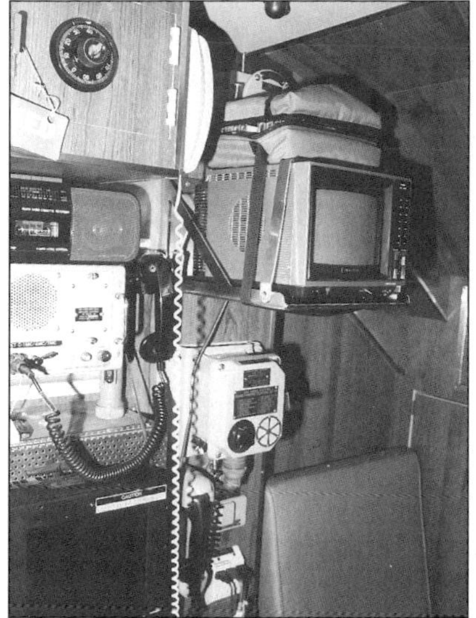

The communictations and recreational equipment in the captain's cabin, USS *Miami*.
(John D. Gresham)

A typical bunk space or "rack" in the forward enlisted berthing area, USS *Miami*.
(John D. Gresham)

The USS *Topeka* (SSN-754) is launched at the Electric Boat Yard at Groton, Connecticut. *(Electric Boat Div., General Dynamics Corp.)*

A Tomahawk surface-to-surface missile is test fired from a submerged sub. *(John D. Gresham)*

The twelve hydraulically operated doors of the *Miami*'s Vertical Launch System for Tomahawk cruise missiles. *(John D. Gresham)*

A Tomahawk cruise missile is launched from the USS *Pittsburgh* (SSN-720) during Operation Desert Storm. *(Official United States Navy Photo)*

Periscope photo of a British frigate taken by Perisher students during their command qualification course. *(U.K. Ministry of Defence)*

HMS *Conqueror* returns home after sinking the Argentine cruiser *General Belgrano* during the 1982 Falklands War. *(U.K. Ministry of Defence)*

HMS *Triumph* is rolled out of the VSEL building barn. *(U.K. Ministry of Defence)*

The crew of a Royal Navy submarine conducts an escape drill. The trainee at left is wearing the Mk8 escape suit. *(U.K. Ministry of Defence)*

A Soviet Victor III nuclear-powered attack sub. *(Official United States Navy Photo)*

A Soviet Oscar-class guided missile sub. *(Official United States Navy Photo)*

A view of the Electric Boat division with the ballistic missile sub USS *Michigan*, SSBN-727, under construction. *(Official United States Navy Photo by William Wickham)*

HMS *Unseen*, one of the new generation of diesel/electric subs operated by the Royal Navy. *(U.K. Ministry of Defence)*

The Royal Navy's newest SSBN, HMS *Vanguard*, is escorted by one of the older "R" class SSBNs. *(U.K. Ministry of Defence)*

USS *Seawolf* (SSN-21) conducts her first at-sea trials on July 3, 1996. *(Official United States Navy Photo)*

Navy and contractor personnel man the underway main control watch, aboard *Seawolf*. *(Official United States Navy Photo)*

A sailor aboard the *Seawolf* operates the fire control tracking system, which uses the latest in rugged touch screen controls. *(Official United States Navy Photo)*

The dining area on the *Seawolf* during chow call. *(Official United States Navy Photo)*

A Russian-built Kilo-class diesel sub, purchased by Iran, is towed by a delivery vehicle in 1996. *(Official United States Navy Photo)*

The converted ballistic missile submarine USS *Kamehameha* (SSN-642) enters Apra Harbor, Guam. The *Kamehameha* was converted to a transport submarine in 1992. She is equipped with two Dry Deck Shelters (DDSs) to support Navy SEALs and SEAL Delivery Vehicles (SDVs) during special operations. *(Official United States Navy Photo)*

A former Soviet Typhoon-class submarine is handled by tugs, on its way to be dismantled. These ballistic missile subs were among the largest ever built. *(Official United States Navy Photo)*

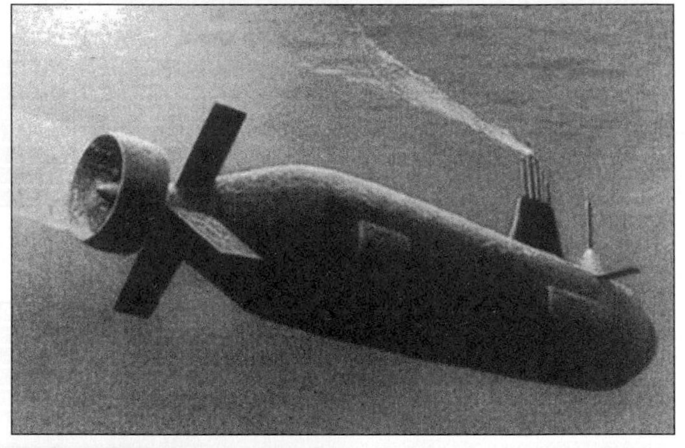

Two artists' concepts for future Royal Navy nuclear submarine designs for the twenty-first century. *(U.K. Ministry of Defence)*

ronment. All the noise from the ice floes breaking apart and grinding together makes it very difficult to locate and track an opposing submarine. In addition the boomer, much like a rat in a warehouse, has a back door to run to under the ice. For this reason, only the most capable of American submarines, an Improved Los Angeles (688I), is suitable.

After a transit to the presumed bastion area, the 688I begins to listen. It maintains a low speed, probably around 5 knots, to optimize the performance of its towed arrays. As the 688I finally enters the target zone, the tracking team in the control room utilizes every sensor and capability of the BSY-1 system to locate and track the opposing boats. This is vital because of the background noises in the ocean (waves, fish, marine mammals, etc.), as well as the noise coming from the ice pack. The first contact is going to have to be a "direct path" contact, so the 688I searches, running in a series of expanding boxes until the first contact is achieved. This contact, which might be either the Typhoon or the Akula, is none too exact with regards to range, but bearing information is enough to continue the hunt. The hunt now becomes a task of patience. The boat will probably go to the quietest routine possible, as the closure to attack might take many hours.

While the American commander probably prefers to avoid the Akula by moving around the Typhoon and using it to mask the 688I's own noise signature, the extensive quieting on the Typhoon will probably preclude this. It simply would be too easy to miss the boomer and stumble into the Akula. Again, patience and stealth is the best tactic of the American boat. The goal at this point is to hold a sonar contact on the Typhoon while trying to avoid the Akula. The key moment comes when a firing solution is finally generated by the BSY-1, hopefully within the CO's designated firing range. Normally it would be helpful to take the time to establish a solid solution to the target to increase the chances of a hit on the first shot. But "polishing the cannonball" with such opponents as the Typhoon and the Akula could cost the chance to get the first shot in. With the tracking capabilities of the Mk 48 ADCAP torpedo, and the danger posed by the Akula, it is now in the best interests of the 688I to "shoot and scoot." As soon as the solution on Typhoon is good enough, the American commander probably orders the launching of a pair of Mk 48 ADCAPs. Each is likely to be launched about 12 degrees off the intercept course (left and right) to the target, so as to cover the entire front 180-degree sector of the 688I. The fish are probably launched in the BSY-1's Short Range Attack (SRA) mode at the high-speed setting, the guidance wires are cut, and the seeker mode is set to active pinging. If he knows the bearing to the Akula, the American commander may choose to fire his other two torpedoes in SRA mode down that bearing also.

With the torpedoes heading on their own toward the Typhoon, the American boat can now run for its own safety (called "clearing datum"). The captain of the 688I is probably going to kick up the speed as fast as possible (over 30 knots), launch some decoys or other countermeasures from the 3-inch signal ejector tubes, and go as deep as the local seabed and the capabilities of the boat will allow. If it's done right, the American boat should have a lead of several miles before one of the Russian boats can launch a torpedo in response. This they will do, though, and the American boat is sure to have one or more Russian torpedoes headed in its direction. But the CIS subs are also running for their lives, kicking out decoys and countermeasures and desperately trying to maneuver out of the way of the oncoming ADCAPs. But with a speed of 60-plus knots and a seeker head that can see targets almost 180 degrees around it, the simple fact is that no submarine afloat can outrun an ADCAP. The encounter now moves to the endgame.

The angling of the torpedoes from the 688I is designed to ensure that at least one of the Mk 48 ADCAPs will "acquire" the Typhoon, though in about two-thirds of the situations, both weapons should track. At this point the Russian boomer is going all out to evade the incoming weapons. It launches countermeasures, trying to jam the seeker heads of the torpedoes and outmaneuver them. This probably will not work. As the Mk 48s close on the target, the Typhoon crew will inevitably hear the pinging of the seeker heads on their own acoustic intercept receivers and know what is coming. At this point the Mk 48's electronic guidance package determines the optimum point for detonating the warhead of each ADCAP on or near the outer hull of the Typhoon and/or Akula. And the effects will be horrendous. If the Akula is hit, it is probably dead. Game over for the enemy SSN. The Russian boomer, on the other hand, will certainly suffer massive outer hull and shock damage. In some cases a breach of the inner hull may occur, causing flooding. Should both Mk 48s hit, they may sink the missile boat immediately. But most likely the massive construction of the Typhoon will allow it to survive the impact of even a pair of ADCAPs. The large space between the inner and outer pressure hulls, as well as the Typhoon's huge reserve of buoyancy (approximately 35 percent of her displacement) will likely allow the boat to survive. If the Russian boat has survived the initial shock and flooding, it may have enough reserve buoyancy to blow its ballast tanks and fight to the surface, assuming it is not under the ice. In any case, with the hull shredded, interior compartments possibly open to the sea, and massive shock damage to the weapons and control systems, it is no longer combat ready. If the boomer has survived, it is making a terrific amount of flow noise, as well as generating mechanical transients (rattling) from the shredded edges of the damaged hull plates beating against each other.

While the torpedo endgame is being conducted, the 688I resumes its quiet routine. In addition, the crew are reloading the torpedo and countermeasures tubes as well as doing anything noisy that they deferred during the approach. Assuming that the American boat has outrun any weapons that were counterfired, it slows and begins the listening game anew. At this point the American SSN commander is faced with a choice. If the missile boat has survived, it will be fighting for its life. And while it is probably incapable of firing its complement of SS-N-20 Seahawk missiles without a major overhaul, the American boat may try to finish the job just to be sure.

And thus the hunt begins again. . . .

Tactical Example—Hunting a Nuclear Attack Submarine

This is a job that has become both easier and harder in recent years. Since 1988 the Russian Navy has voluntarily retired a whole generation of its submarines. Many, perhaps all of the Hotel, Echo, and November classes of SSNs are reported to have been deactivated—in some cases hauled out of the water to rot while Russian naval officers seek the advice of their American counterparts on the best way to dispose of their still "hot" reactor plants. Early Victor-class SSNs have reportedly been offered for sale to the West as ASW adversaries (the U.S. Army's National Training Center has a large supply of Soviet-made fighting vehicles, obtained through less conventional means). The Russian Navy appears to be reverting to its entirely legitimate role as its country's maritime defense force while that country's land forces continue to assume their place as that country's principal defense arm. That means a smaller Russian Navy, and one that remains closer to home.

But nuclear submarines are not necessarily creatures of one's home coast, and those Russian SSNs remaining in service are the best ever produced by that country. The Victor III is the mechanical equivalent of the American 637 (Sturgeon) class, and the Akula (the Russian word for *shark*, applied to that class by NATO after the West ran out of letter-code designators) is reportedly equivalent to an early 688 (Los Angeles) class. They are, in a word, close enough in performance to Western SSNs that the skill of the captain and crew becomes the deciding factor.

Tracking one of these submarines takes on the aspect of a one-on-one sporting event. And it is an event that has been played out many times since both sides acquired SSNs in the early 1960s. During this time, the Soviet forces were making preparations for a possible ground war against NATO in western Europe. And much as the German U-boat fleet did in World War II, the Soviet Navy was planning to support them with a massive surge of SSNs and SSGNs (Nuclear-Guided Missile Submarines) into the North Atlantic to stop convoys with reinforcements from reaching the NATO forces. Since proficiency in such skills takes practice, the Soviets began to have their SSNs make regular patrols into the Atlantic Ocean and near to the American coast. Usually these were conducted by newer boats such as Victor IIIs.

Part of the problem with staying ahead of the Soviets in those days was recognizing when they were using or doing something new. During the 1980s the Russians brought out a large number of new nuclear submarine classes, and early identification and classification was a top priority for the boats of the various NATO powers. Usually this was accomplished by a boat sitting at a "gatekeeper" station off Petropavlovsk and Vladivostok (in the Pacific), and off the Kola Peninsula near Murmansk and Severodvinsk. The job of the gatekeeper was to sit and watch. Anything that went in or came out was carefully noted and catalogued. Occasionally the sub would stick an ESM/Comint mast up and sniff the air for the electronic emissions that are part of every military base in the world.

There is a story, told in whispers and with guarded glances, about one of the greatest of the gatekeeper boats and her skipper. It is only a story, and neither the U.S. Navy or Royal Navy will officially state that it ever took place, but such are the stories that come from the silent service.

Sometime in the mid-1980s a gatekeeper boat was off the Kola Inlet, doing its job day after day. The sonar watch detected a submarine coming out of the barn from Severodvinsk. When the noise signature of the power plant and the other machinery on board did not match any known class of Russian boat, the captain of the U.S. boat decided to trail it and learn all he could about this new machine. Perhaps it was the first of the Sierra- or Oscar-class boats, or even the one-of-a-kind Mike-class boat with its titanium hull and liquid sodium reactor. Whatever it was, though, the U.S. commander was intent on getting to know everything possible about the new Soviet sub. The U.S. skipper carefully and quietly started stalking the Russian boat, probably from the rear, at a short distance.

In the chase that followed, the American sub listened and watched every move of the new boat. The sounds of the propellers and the all-important blade rate, which is used to calculate the speed of a ship or submarine. All of the machinery noise from the reactor (or reactors—many

Russian boats have two), turbines, and pumps. They may even have heard some of the day-to-day living noises aboard the Soviet boat. The bilge tanks being pumped out, the TDU dumping garbage, and maybe even the sounds of hatches closing and pots and pans clanging in the galley. And through it all, the American boat and her crew remained undetected by the Russian boat and any supporting vessels that might have accompanied her.

After a period of time—and here the story begins to take on the air of unreality that is a hallmark of the *true* submarine stories—the Soviet sub came to the surface and slowed down. As the American sonar crews observed the Russians going to the surface, the American skipper apparently decided to try for the grand slam of submarine intelligence-gathering coups, getting some hull shots of the new Russian boat (video pictures of the hull, propellers, and control devices beneath the surface).

Such an operation is done by running underneath the target boat, raising the periscope equipped with a low-light video camera, and running a pattern around the hull to collect the video pictures. This is so difficult and dangerous that captains of U.S. submarines are almost never ordered to try it, as bumping a target can be non–career enhancing. On the other hand, successfully gathering hull shots is a sure sign that the boat's skipper has the right stuff and is worthy of promotion to higher command. And with only a few O-6 (captain) command slots available for boomers, tenders, and squadrons, the competition is fierce among the various attack boat skippers.

What happened next was a marvel of seamanship. The American boat was able to make at least one (several tellers of the story say more) pass around the Russian sub's undersides, and not once get noticed! Up and down the sides of the Soviet boat, the American skipper drove his periscope, obtaining the broadest possible coverage of the target. The coverage apparently included the control surfaces, propellers, and several sonar arrays. The quality of the video pictures was excellent, adding much to NATO's understanding of the new Russian boat. And maybe most impressive of all, she was able to back away, continue the chase, and eventually resume her gatekeeper position off the Kola Inlet.

The achievement was so impressive in its day, so the story goes, that the skipper was awarded a "black" Distinguished Service Cross (i.e., the recipient is unable to wear it, but the decoration appears in his service file folder or "jacket"). While such peacetime decorations are not unprecedented, they are extremely unusual, and the award of such a thing would be an indication of how important the U.S. high command considered the action.

This is the way the game of hide-and-seek went for almost forty years between the nuclear boats of the United States and the Soviet Union. And the game continues today. Only recently, there was a *very* public airing of a minor collision between the USS *Baton Rouge* (SSN-689) and a Russian Sierra I north of the Kola Inlet. There were some bent hull plates, some exchanges of diplomatic messages, and minor apologies between the United States and the Russians. But have no doubt, the day-and-night stalking still continues as this book goes to press.

Tactical Example—Escorting a Boomer

In World War II the U.S. 8th Air Force found out the hard way about the price of running bombing raids into Germany without fighter escort. The big, heavy bombers were no match for the quick, heavily armed fighters of General Adolf Galland's *Luftwaffe* fighter command. Thus it was no surprise that as soon as they could be obtained, the 8th Air Force started to deploy fighters to escort the bombers against the danger the *Luftwaffe* fighters posed. These fighters not only reduced bomber losses but also tore out the heart of the *Luftwaffe* fighter command, making the invasion of Europe possible and victory that much easier.

Today, lessons such as these have not been lost on the operators of the boomer force in the U.S. Navy. The Ohio-class SSBNs are the largest and most capable FBM boats ever deployed by the United States, and also the most valuable. The Navy is proud of saying that no U.S. SSBN has ever been tracked while on patrol. But what about when it is headed out to patrol? With so many of America's strategic "eggs" in just a few Ohio-class hulls, they clearly are crown jewels needing protection. And when the boomers come out of Kings Bay or Bangor, they are extremely easy to see, whether by satellite or just a set of human eyes watching as they steam up the channel. Once they are at sea they fade away into the depths, but while departing and arriving at the base, they are vulnerable.

While the U.S. Navy has never made a big deal about such things—and with the end of the Cold War it is unlikely that they ever will—such vulnerability is a concern when you have only a few of the big Ohios to carry over 50 percent of America's total nuclear weapons load. All it would take is some easy cueing from a source ashore to tell an enemy submarine just when a boomer might be headed to sea. Thus it makes good sense to have the big FBM boat escorted out to sea by attack submarines, much as a fighter might escort a bomber on a bombing raid. It should be emphasized that a hostile boat would probably not try to get a shot in, though in wartime conditions this is always a possibility. More likely the threat boat would try to get on the tail of the Ohio and track it for as long as it could.

Let's suppose someone wanted to try tracking an Ohio as it came out of the channel at Kings Bay, Georgia. The continental shelf near Kings Bay is somewhat longer and flatter than at Bangor (the seabed drops right to the continental slope at the mouth of Puget Sound), giving a potential enemy submarine a somewhat easier time finding the Ohio as she comes out. Some time before the boomer is scheduled to leave port, one of our SSNs, probably a Los Angeles–class boat, will be stationed off the mouth of the channel to sit and watch for any signs of foreign submarines. The U.S. boat's mission will be to sanitize the area, making sure no other submarines have covertly entered American territorial waters to lie in wait for the SSBN. It will be a long, boring process, with many of the same kinds of problems described in the previous hunting scenarios. They will slowly patrol the area and listen, looking for any sign of something unusual or man-made.

If they find another submarine at this time, it will be reported quickly, and action will be decided upon by higher authority. More likely, though, is the scenario in which a hostile submarine is waiting just outside the twelve-mile limit of American territorial waters. In this case the Los Angeles will probably try to sit astride the planned route of the Ohio and wait for any sign of activity. If such contact occurs, the action that follows might go something like this:

As the Ohio comes out (escorted by support and security vessels to keep, if nothing else, the Greenpeace protesters at a safe distance) and prepares to dive, the Los Angeles continues its job of sanitizing the ocean ahead of the boomer. Much like a sheepdog herding a flock, its job is to interpose itself between the SSBN and any threat until the boomer can slip quietly into the deep waters off the Carolina/Georgia coast. Once an Ohio is free of the continental shelf, even the latest 688I-class SSN would find it almost impossible to track.

The Los Angeles continues ahead of the boomer, until it gets the first "sniff" of a hostile sub. Then the engagement takes on all the aspects of a game of chicken with tractor-trailer trucks. The Los Angeles closes with the threat boat, trying to get it away from the Ohio with everything short of actually ramming it or firing weapons. The SSN initiates maneuvers conforming to the rules of the road, which require the hostile boat to evade. The American SSN might launch noisemakers and other countermeasures in an attempt to make so much noise that the Ohio will be lost in the background. Another technique has the Los Angeles masking the Ohio by standing along the path between it and the hostile sub, and blasting away with its spherical sonar array as a jammer.

If the threat sub proves to be particularly obnoxious, the American skipper might even engage in a maneuver to force the hostile boat's skipper either to take evasive action or suffer the possible damage and embarrassment of an underwater fender bender. Whatever maneuvers the attack boat chooses, the desired result is that by now the Ohio has slipped into the deep waters off the continental shelf and is silently on the way to her designated patrol area. Once this is accomplished, the Los Angeles probably breaks off the chase and heads for home.

Thus begins another in the more than 3,000 FBM patrols that the United States has run over the last three decades. The SSN will have helped make it a successful one, that is, one in which the boomer returns to base with all twenty-four of its missile tubes still loaded, missiles unfired. Some might claim that the above scenario is only the wildest speculation and conjecture, and perhaps this is true. But just what was that Victor III that surfaced off the

Carolina coast in 1983 doing there? Just remember that the submarine bases at Charleston, South Carolina, and Kings Bay, Georgia, are right in that neighborhood. Do you think the Victor was there just to photograph the resort at Hilton Head? Hardly.

Tactical Example—Hunting a Diesel Submarine

One of the few growth industries in the defense world today is the diesel-electric submarine market. Since the end of the Cold War, more and more small- to mid-size navies have seen these compact, cost-effective craft as a way to make up for whatever protection they may have enjoyed from whichever side they allied themselves with during the Cold War. Unfortunately, because of cutbacks in the defense industry worldwide, some of the nations that produce such boats have sold their wares to nations that the rest of the world might consider somewhat less than responsible. China, India, Pakistan, Iran, and Algeria are just a few of the countries that have decided to invest heavily in diesel boats.

Surely the Volkswagen of the current generation of diesel boats is the Kilo-class boat produced by the CIS/Russia. This trim little boat is compact, has a good combat system, adequate weapons and sensors, and is *very* quiet. This makes it an excellent candidate for operations in straits and other choke points. In addition, a well-handled Kilo is almost impossible to detect passively when she is running on her batteries. And so our little story begins.

Let us suppose that the Islamic fundamentalist movement takes a serious hold in Algeria, along the coast of North Africa. And let us again suppose that the local ayatollah decides the merchant traffic passing along his coast should have to pay some duty for the privilege. It might then be possible that the Algerian Navy, the recent recipient of several Kilo-class boats, will be ordered to give the western merchants a demonstration of what might happen if they do not comply with the wishes of the new Islamic government.

An ideal way would be to seal the nearest choke point, then try to collect reparations to refrain from doing it again. For a cash-starved country like Algeria, this toll might be considered an excellent way to generate capital. The likely place for this demonstration would be the Straits of Gibraltar. Not only is it an ideal place for a diesel boat to operate, but the symbolism of doing it under the nose of the British Empire would be tough to resist.

The first notice of what was happening would probably be the "flaming datum" of an exploding merchant ship. Most modern torpedoes are designed to explode under the keel of the target ship, snapping it in two. If this were done to a tanker, for example, there would likely be a massive oil spill and fire, as well as wreckage that might float as a hazard to navigation for some time. This, combined with the inevitable declaration from the Algerian government, would undoubtedly cause a reaction from the Western powers. For hundreds of years Great Britain has held control of the seas around Gibraltar, and any mischief in the area would probably make them want to deal with it themselves. The likely candidate for this ASW extermination job would be a Trafalgar-class nuclear boat, because of its ability to deploy rapidly to the area threatened by the Algerian Kilo. Most folks do not realize that a diesel boat is actually just a mobile minefield. It simply does not have the strategic mobility or sustained speed of a nuclear boat, a simple fact that is lost on critics of nuclear submarines.

The deploying T-boat is likely to have some help in the form of RAF Nimrod ASW aircraft. In addition, it is a safe bet that the British have seeded the straits with a variety of acoustic sensors, and the area is about as wired as a pinball machine. The problem for the British hunters is the adverse noise conditions in the straits. There are several thermal layers, which make passive sonar almost useless. In addition several currents, overlapping and opposed in direction, generate a lot of flow noise. All in all, the Straits of Gibraltar is a miserable place for passive ASW hunting.

Fortunately, though, the nuclear submarine has another advantage over the diesel boat besides sheer mobility. That advantage is the huge active sonar array positioned in the bulbous bow of the boat, which is able to send out pulses of sound and bounce them off a target submarine. A special operating mode makes it even more effective: in areas with relatively flat, hard bottoms, a technique called "bottom bounce" can be used. Much like skipping a stone across the water, an active sonar can bounce sound waves off the bottom to contact another submarine. Using this technique, a nuclear submarine might contact an almost-silent diesel boat at ranges beyond 10,000 yards. And as an added benefit, because of all the reverberations from the sound waves bouncing off the seabed, the target submarine probably will not be able to tell what direction the active signal is coming from.

The Trafalgar enters the straits from the Atlantic side. The British may try to use their other assets, the Nimrods in particular, to help drive the Kilo into the hunting Trafalgar. The Nimrods may be tasked to drop active sonobuoys. These, combined with active sonars from ASW helicopters, might just make the Kilo captain move deeper into the straits, right into the waiting T-boat. The aircraft, however, will not be allowed to drop any ASW ordnance on it. With many submarines of various nations traveling through the straits, and the closeness of one of their own nuclear boats, the possibilities for a "blue-on-blue" or friendly fire confrontation are simply too high. The Trafalgar is like a surgeon's scalpel compared to the bludgeons of the aircraft.

Once the British think the T-boat is within range of a bottom bounce detection, the Trafalgar would probably use her 2020 active sonar to scan for the Kilo. This will be extremely disconcerting for the Kilo captain, with the buoys and active sonars of the aircraft and helos driving him from the Mediterranean side, and the blasting from the active sonar of the Trafalgar. He may choose to find a shallow spot and bottom his boat in an attempt to wait the British forces out. This will not work. With the on-station loiter time granted by its nuclear power plant, the Kilo will be out of battery power and supplies to run her environmental control systems long before the beer runs out in the wardrooms of the T-boat.

Inevitably the Kilo will have to make a run for it, and that's the time for the kill. The advantage of active sonar is that range and bearing to the target are known with a fair degree of accuracy. An added bonus with this powerful generation of active sonars is that the acoustic intercept receiver on the Kilo will be so swamped with noise (like a stereo system with the volume too high—you cannot make out any discrete sound), they will not hear anything but the sound of the British 2020 sonar blasting away. Once the T-boat has closed to the desired range (probably over 10,000 yards), it is time to prosecute the Kilo. The Trafalgar may launch a pair of Spearfish torpedoes in high-speed mode, active pinging, with the wires acting as data links to the weapons.

The Kilo is likely to hear nothing of this. Only when the seeker heads of the Spearfish have acquired the Kilo will the active sonar of the T-boat be secured, and then the crew of the Kilo will hear over their acoustic intercept receiver the pinging of two Spearfish torpedoes already commencing their endgames. Unlike the previous scenarios, in which the nuclear boats could sometimes run from torpedoes and possibly outmaneuver them, the Kilo just does not have that option. Its relatively slow speed makes it something of a sitting duck, and the end will come quickly. This time there will be no doubt, for when the first torpedo hits, it will kill the little diesel boat and all its crew. In all likelihood, all that will be left is scrap metal and fish food.

And that's the way to deal with modern Barbary Pirates.

Tactical Example—Battle Group Escort

The big gun of the fleet is still the aircraft carrier battle group (CVBG), which for that very reason is itself a target. The carrier remains the best platform for projecting power from sea to land, and the best for establishing *presence*, a term that means just what it says. A carrier and her battle group can appear on the horizon and just *be there*. As a police car can calm a neighborhood merely by cruising down the street, so can a powerful air/surface force let people on land know that someone cares what is happening.

The most likely threat to a carrier is a submarine armed with antiship cruise missiles (SSMs). Though unlikely to cause fatal damage to a supercarrier, a few well-placed SSMs can force her to leave the scene of action for repairs. The range of modern cruise missiles (up to 300 miles) makes the task of protecting the carrier far more complex than it was only two decades ago. Another problem is the decreasing number of ASW escorts available to the commanders of CVBGs. In just the last couple of years the U.S. Navy has retired dozens of cruisers, destroyers, and frigates. Since the submarine remains the primary threat, another submarine must be one of the protectors.

The most formidable dedicated cruise-missile submarine (SSGN) is the Russian Oscar class (nicknamed "Mongo" by some NATO submariners because of its awesome size). The Oscar-class SSGN is, in some ways, the Russians' first modern submarine. It is large and relatively quiet (much like a Sierra-class SSN) and is equipped to stream a large towed-array sonar. This boat, designed specifically to be a carrier hunter, is equipped with twenty-four SS-N-19 Shipwreck SSMs as well as a full array of torpedoes. It is the single most powerful attack submarine in the world, and thus must be hunted by the best boats we have, the 688Is.

Currently each CVBG usually has a pair of SSNs assigned to provide long-range ASW protection. Unlike the surface escorts, which have to stay within a few dozen miles of each other, the subs may be hundreds of miles from the main group. They will likely operate in clearly defined ASW kill zones, into which only they are allowed to operate and shoot. This is designed to minimize the chances of a "blue-on-blue" ASW encounter.

Hunting SSGNs is a most interesting game, different from other ASW tasks. Unlike SSBNs, which run silent and deep, the CVBG relies on mobility for its defense. And when the carrier moves swiftly, so must the hunting SSGN. Speed reveals any submarine's vulnerability. Speed creates noise and degrades sensor performance. The SSNs tasked to defend the carrier know both where and how fast the battle group is going, and can position themselves in ambush for whatever missile-carrying hunter may be listening. In addition, the American force may have the edge of a Surveillance Towed Array System (Surtass) ship supporting the CVBG. Using an advanced towed array, the Surtass ships are like mobile SOSUS listening posts, and the data collected can be forwarded to the CVBG commander and the hunting SSNs.

The pattern of this hunt will be sprint-and-drift. The hunters on both sides alternately race forward, then slow down to listen. As in all undersea encounters, the side that can hear first and farthest away has the biggest advantage. Knowing where and when an SSGN would have to approach, the U.S. sub has the ability to stay quiet and wait for the Oscar to come to it. Because of its need to obtain targeting data from the Russian RORSAT, the Oscar has to come shallow periodically to raise its satellite data link masts. This causes hull popping and mast flow noises. Thus it is entirely likely that the 688I can be guided by targeting updates via the ELF/VLF radio circuits to a point where it will be able to obtain a direct path passive sonar contact to the Oscar. This will probably occur at a distance of 10,000 to 16,000 yards.

As in the hunt for the Typhoon, the 688I must go to an extremely quiet operating routine, to remain undetected by the towed array of the Oscar. But unlike the hunt for the boomer, here time is of the essence. Potentially, the Oscar can fire its missiles once it is within range of the CVBG. This means it must be eliminated quickly and effectively. The U.S. skipper is likely to try maneuvering to a position behind the Oscar, so that any torpedo hit will strike near the propeller shafts. This is likely to pop the shaft seals, flooding the engine room of the Russian boat and hopefully sinking it. All the while the fire control technicians operating the BSY-1 system will be "polishing the cannonball" on the firing solution to the Oscar. At 6,000 to 8,000 yards, assuming the Oscar has not yet heard them, the U.S. skipper may launch a pair of wire-guided Mk 48 ADCAPs. These are fired initially in the slow-speed setting, using the wires to guide the weapons and provide data back to the U.S. boats. The fire control technicians may even try to "swim" the weapons under a thermal layer to mask their noise signature from the sensors of the Oscar.

Inevitably though, the Oscar hears the two Mk 48s and begins to react. It counterfires torpedoes down the bearing of the attacking Mk 48s, forcing the 688I skipper to cut the guidance wires and run for cover. Its distance lead over the Russian fish, as well as efficient maneuvering of decoys, should allow the American boat to survive. The same may not be true of the Oscar. The captain of the Russian boat tries the same evasion tactics as his American opponent, but they are probably not as effective.

As in the Typhoon example, at least one and possibly both ADCAPs are likely to hit their target. And if the desired shaft hits have occurred, then the Oscar is dead in the water. Even if only a single hit has been made, the 688I has accomplished its mission. The Oscar is badly hurt, and likely suffering from severe shock damage. It may even have to surface. In any case, it will be making horrendous amounts of flow noise and mechanical transients. The U.S. skipper may reattack and finish off the Oscar, or he may also call the carrier to give it the coordinates of the damaged missile boat. Within a very short time the carrier group could have a flock of S-3B Viking ASW aircraft and SH-60 ASW helos over the damaged Russian boat to finish it off. Much like a wounded bear being stung to death by a swarm of bees, it would die. And the American boat can now head out on another mission.

MISSION #2—ANTISURFACE WARFARE

The nineteenth-century French Jeune Ecole first codified the idea that a navy isn't the real target of maritime warfare—the real target is what navies were designed to safeguard, merchant shipping. The sea is, before all things, a highway over which nations trade. And navies were invented to protect that, first from pirates who were little more than thieves at sea, and then from foreign navies whose thievery was on a somewhat grander scale. One might say that the real role for the submarine grew from this doctrine. The first submarines were too slow to be really effective at hunting other warships but quite fast enough to seek out and kill the slower and more fragile merchant tubs that carried the things nations need: food, raw materials, manufactured goods. Since the global economy has made all countries into island nations surrounded by water, the vulnerability of international maritime trade is made greater still by the fewer, slower, larger, and massively expensive merchant vessels of today. The environmental consequences from even minor damage to a single large crude oil carrier represent yet another way in which the world as a whole can be at risk. Navies exist to protect the trade and the traders, and a threat to either is a threat to both.

Tactical Example—Holding a Choke Point (Interdiction of a Surface Action Group)

The simplest example of a choke point is a highway intersection, a relatively small area through which people from distant places must pass on their separate journeys. Just as the intersection is a convenience for merchants who build shopping centers and people who establish maritime trading centers, so it creates highly rewarding hunting grounds. In World War II, the first Japanese task group to be detected was preparing for the invasion of the Kra Peninsula and the subsequent descent on Singapore, which guards the Strait of Malacca. England spent a great deal of its history seizing and building upon such places as these, even before Alfred Thayer Mahan published his thoughts on their importance. The Falkland Islands became British property because they are conveniently close to the Straits of Magellan. Ascension Island is in the middle of the Atlantic Narrows. Malta lies close to the Straits of Sicily. Gibraltar sits on the entrance to the Mediterranean. Such was the vision of the English in the days of sail.

Ships move faster now, but the choke points remain. In these places that people must pass through, arrival time and engagement range are predictable quantities.

A lucky submarine will hug a shallow bottom. Shallow water, not uncommon in straits, generally makes life easier for a submarine, though SSNs usually like at least 600 feet/200 meters of water to operate. Given time, the submarine will sniff around, learning currents and environmental factors. The mouth of the Mediterranean is known for the treacherous mixture of warm currents and cold, making for confused sonar conditions. In other places such conditions might mitigate against a sub, but long-range detection is of less importance when you are already astride the place where others must pass.

The other side knows this, too, of course. The mere possibility that someone might be there to threaten your battle group or your crude carrier forces you to take this threat seriously. Willie Sutton robbed banks because, he said, "That's where the money is." Choke points are where the targets are. You can bank on it.

Let us consider the most famous submarine action in recent history: the sinking of the Argentine cruiser *General Belgrano* during the 1982 Falklands War. Before the British task force entered hostile waters the Royal Navy deployed a trio of nuclear submarines along the most likely approach routes to the islands. Because of his limited air power and surface-to-surface missile capabilities, Admiral "Sandy" Woodward was counting on this trio of boats to be the flank guards against any type of counterattack by the Argentine Navy. As it turned out, they were the only units of the Royal Navy to engage the major surface units of Argentina during the war.

In the last few days of April 1982 the Argentinean surface fleet was split into three task groups. Their plan appears to have been based on a three-pronged pincers movement against the British task force from the north, south, and west. The northern group was composed of their aircraft carrier *Veinticinco de Mayo* (Twenty-fifth of May) with a small air wing of A-4 Skyhawk attack aircraft, and several guided missile destroyers carrying Exocet SSMs. The western group was composed of several Exocet-armed frigates. The southern group was potentially the most dangerous force of all, composed of the cruiser *General Belgrano* (the former USS *Phoenix*) armed with 6-inch guns, Exocet SSMs, and Seacat SAMs, accompanied by two Exocet-armed destroyers of World War II vintage.

It is likely that national intelligence sources of the United Kingdom and their allies noticed the planned movement even before the ships raised anchor and left their harbors. And once they had sortied, it must have been fairly straightforward for the Royal Navy operations center (known as HMS *Warrior*) at Northwood, England, to feed the updates to the subs via their satellite communications. The three British boats were placed along the three surface groups' lines of advance, and were left in waiting while the British naval and civilian leaders decided whether to shoot.

The key question here was whether or not the Argentine forces would attempt to penetrate the 200-mile-radius total exclusion zone (TEZ) from Port Stanley in the Falklands. Clearly if they tried to penetrate it, there would be no question but to attack with the nuclear boats. But the ships did not come on directly and seemed to rally just outside the zone, though quite close enough to dash in at a moment's notice. The northern group was trying to find some wind to launch her strike of A-4s at the British task group, though (amazingly for the weather in the South Atlantic) it was calm and windless.

The choke point in that area was, ironically, not a strait but the extremely shallow water. The Argentinean southern group was operating over a shallow rise in the ocean called the Burdwood Bank, which made difficult operating conditions for HMS *Conqueror* (S-48), the southern boat in the British barrier. This hydrographic choke point was a major problem for the British SSN, and became a factor in the decision coming from 10 Downing Street. Already the *Conqueror* and another boat were tracking their assigned target groups, and a decision was needed from on high.

Late on May 2, 1982, the message was sent from Northwood authorizing the sinking of the *Belgrano*, and any of her escorts that attempted to intervene. Even though it was still some distance outside the TEZ, the *Conqueror* was the first to strike. Her captain, Commander Christopher Wreford-Brown, set up a classic Perisher approach on the *General Belgrano*. Loaded in his five torpedo tubes were three World War II–vintage Mark 8 torpedoes and a pair of Tigerfish Mod 1s. The plan was to use the Mk 8s first because of their larger warheads (800 lb/363 kg versus 200 lb/91 kg for the Tigerfish), and save the Tigerfish for a second shot if required. If the Mk 8s worked on the first try, the Tigerfish would be available for a shot or two at the escorting destroyers if necessary.

In the plotting area of *Conqueror*, Lieutenant John T. Powis, the boat's navigator, carefully plotted the intercept from ranges and bearings called by the commander on the periscope, and inputs from the sound room. It was a totally normal approach, which later would be judged considerably easier than most of the approaches made during a Perisher course. Wreford-Brown maneuvered the *Conqueror* just 1,200 yards off the projected track of the *General Belgrano* and patiently waited. The Argentine ships continued blindly along their track, completely oblivious of the coming danger. And then it was time.

Just before 1600 hours on May 2, 1982, the only combat torpedo shots ever fired by a nuclear submarine were launched from *Conqueror*. The three Mk 8s were angled in a way designed to ensure that at least two of them would hit the *General Belgrano*. And that is exactly what happened. The first Mk 8 hit forward near the bow, tearing it from the ship. The second one struck in the engineering spaces, causing a complete loss of power and massive flooding. The *General Belgrano* immediately took on a heavy list to port, and within minutes it was sinking. Her captain had no choice but to abandon ship and have his crew take to the life rafts. (Ironically, the exact same type of hits in the same places had sunk her sister ship, the USS *Helena*, during the Battle of Kula Gulf in 1943.) Some 400 of the *Belgrano*'s crew of over 1,000 perished in the sinking and while waiting to be rescued.

In addition to the two hits on the cruiser, the third Mk 8 appears to have hit one of the escorting destroyers, though it failed to detonate. Unfortunately for the crew of the *Gen-*

eral Belgrano, the escorting destroyers did not even know what had happened until they looked and noticed that the cruiser was no longer in formation. It would be almost forty-eight hours until all the survivors of the sunken cruiser were finally rescued.

Aboard the *Conqueror*, there was the satisfaction of hearing the sounds of two solid hits and the breaking-up noises from the cruiser. In addition, the sub reported the dropping of a few depth charges, though this has never been confirmed by the Argentines. Since the escorting destroyers had just continued blindly on, there had been no opportunity to immediately follow up the first attack. And when they moved away from the TEZ, it is likely that the prevailing rules of engagement prevented further action. The *Conqueror* continued on station, as assigned.

As for the Argentine Navy, the effects were rapid and enormous. One story, told at the bars where submarine officers go after hours, says that as soon as the carrier group to the north of the Falklands got word on the sinking, they immediately reversed course and headed back to port. The story goes on to say that apparently this spoiled the opportunity of another Royal Navy submarine commander, who allegedly was watching the oncoming carrier group through his periscope. Royal Navy gossip has it that he watched the carrier and her escorts turn for home less than thirty minutes before he himself would have gotten in his shots at the *Veinticinco de Mayo*. The Argentine surface forces never again ventured out of port during the war, and the British had, in essence, reduced the conflict to a set-piece battle against the air and land forces of Argentina. All for the expenditure of three World War II–vintage torpedoes, perhaps the most cost-effective naval victory in history.

Tactical Example—Maritime Interdiction (Attack of a Convoy/Amphibious Group)

This is a high-risk situation for everyone. A convoy by definition is a large group of valuable ships protected by a force of warships. If your enemy has a convoy, he is transporting something important to his war effort, something that you don't want to let arrive. An amphibious warfare group is a little different; in this case the cargo is the most precious of all for the enemy and the most dangerous of all for you: fully equipped combat troops who have a job to do. You've got to try to stop the enemy in either case, while he must guard a moving asset. In fact, that asset will be moving as rapidly as possible to minimize the risk—the faster they move, the less time you have to attack. But unlike a carrier group, in which every ship has the ability to defend itself, most of the flock in a convoy is relatively helpless. For the amphibious group, there is one other difficulty. While a convoy probably wants to go from one friendly port to another, the amphibs by definition are heading into harm's way—they want to go where you and your allies live. That means the submarines will have to hunt their enemy on the enemy's ground.

Let's suppose that the government of the Ukraine has decided to support some of their former allies in the Balkans, the Serbians perhaps, with an amphibious expedition from the Black Sea to the Adriatic. It might be composed of some six to eight ex-Soviet Ropucha- or Polnocny-class landing ships with a regiment of landing troops aboard. The landing ships probably have an escort of four to six frigates (like Krivaks or Grishas) and/or corvettes (like Pauks or Tarantuls), though nothing like the kind of escort that the old Soviet Navy used to be able to put together. In any case, this is the very kind of intervention that the UN is desperately trying to avoid in a festering part of the world. While NATO air and surface forces could certainly deal with such a group, it would be messy. And there would be repercussions: further confrontations between east and west might develop. Or, someone could just deal with it. Someone who has something that can just make things disappear.

Word of the expedition would not be difficult to find out. States like the Ukraine are filled with opposition groups, and the national intelligence assets of the United States clearly notice the gathering of the ships and the movements of the troops and vehicles to the port. Thus the United States would have several days to coordinate the necessary assets and move a 688I into the Aegean or Adriatic to intercept the amphibious group.

When it comes out, the amphib group has the landing ships (say eight of them) in two columns, with a circle of ASW escorts (say four of these) surrounding them at the flanks. The key problem for the U.S. skipper is to do enough damage to stop the group but not necessarily kill all the troops on the landing ships. One way to do this is to destroy the escorts in plain sight of the landing ships, so that they will realize how naked and vulnerable they are, and go back home. And this is exactly what the American boat decides to do.

The one constraint is that the U.S. skipper must make sure that the weapons used are not unique. This is to say, a torpedo is a torpedo, a Harpoon missile is a Harpoon missile. Many nations have these things. Using these weapons would not leave a "smoking gun" pointing at the United States. "Credible deniability," they call it. But using a unique weapon like a Tomahawk antiship missile would point the finger directly at the United States, so these powerful weapons simply will not do.

The most favorable angle of fire is directly down the amphibious group's line of advance. Since the best ASW ships in the escort will probably be out front, these will be the first targets. The favored weapons would be two pairs of Mk 48 ADCAPs, each pair being controlled by a fire control technician at the BSY-1 consoles in the control room. In this way, the only thing the oncoming escorts will hear is the high-speed sound of torpedoes. There will be no way to know who manufactured them, or who fired.

The approach may be aided by targeting assets like a P-3 Orion patrol aircraft or other over-the-horizon targeting systems. Every now and again the 688I pokes its communications mast above the water for a short time, takes in the latest tracking data on the amphibious group, and then goes back to the job of positioning itself along the group's line of advance. Eventually the BSY-1 system begins to pick up indications of the oncoming vessels. The first contacts may be "convergence zone" (CZ) contacts, which occur at regular intervals of about thirty miles from the target. In this way a submarine can hear a surface vessel at something like ninety miles, or the third CZ. But most likely the noisy diesel engines of the landing ships will allow the U.S. boat to hear them coming from over a hundred miles away.

By now the boat is at its most quiet routine, so that the oncoming escort vessels, as well as any ASW aircraft, will not be tipped to the presence of the intruder. Now the game becomes one of patience, staying quiet while the Ukrainian force bears down. Finally the last of the CZ contacts die out, and the first direct path contacts begin to be heard. The captain of the U.S. boat now tries to place the boat right down the middle of the group's course track and waits for them to close. When the range gets down to about

15,000 or 20,000 yards, the time for action has arrived. The four ADCAPs are launched in the slow-speed mode and guided under any thermal layer that might be present, so their passage to the two leading escorts will be as covert as possible.

Even when the torpedoes get closer to their targets, it is unlikely that the escorts will finally hear them and react. Now is the time to move the Mk 48s up to high speed (60-plus knots) and run them right into their targets. There will be little for the targets to do. With a top speed of around 30 knots, the escorts won't be able to outrun the fish anyway, and with the wires still guiding them (each ADCAP has ten miles of the stuff, remember), it should be an easy matter to guide the torpedoes under their targets and detonate them. The effects will be incredible. A single Mk 48 detonated under the keel of a frigate will, at the minimum, snap it in two.

At this point the next move is up to the senior Ukrainian officer present. If he is smart, he will turn around and run for port. If he is stupid, he will attempt to charge his remaining escorts into the area, call for some air support if any is available, and try to find the intruding boat. By this time the American skipper has reloaded his tubes and is setting up shots on the two remaining escorts. This will likely lead to the destruction of those ships as well. Should this happen, the captains of the landing ships will undoubtedly have the sense to run for home. The Ukrainian adventure is over. If the Ukrainian government is smart, they will not even bring up the fact that the incident took place.

As for the American skipper and his boat, their only problem is slipping quietly and discreetly away. And this they will do. . . .

Mission #3—Covert Missions/Special Operations Support

Into a world that is moving away from major war and toward a long-hoped-for global peace comes a new and intermediate hazard: low-intensity warfare. Actually, this is not a new phenomenon. It used to be called banditry, brigandage, or other desultory names by professional soldiers—when a soldier dies in such a conflict, he's just as dead as one killed on Normandy Beach. As a former commandant of the Marine Corps put it, "If they're shooting at me, it's a high-intensity conflict." That said, however, the rules are a little different. Today one must be more circumspect.

The new reality of warfare is a modification of the old. What was once reconnaissance becomes covert operations, putting small teams of exquisitely trained specialists into a place where they ought not to be, allowing them to do their job, whatever it may be, and then getting them out. If the job is done right, nobody will ever know who did it; and in many cases, nobody will ever know what was done at all.

Accomplishing something like that means stealth, and stealth is the submarine's stock in trade.

Tactical Example—Special Operations Insertion and Extraction

The quintessential special ops mission: pictures that need to be taken, an asset (human or electronic) that needs to be recovered, a bridge that needs to be rearranged. Whatever the particulars, it is essential that the mission be carried out. Such things are, by definition, outside the scope of normal national intelligence assets and may be considered to be acts of desperation. Thus they must be undertaken by personnel who have no desperation in their souls—in short, submariners and SEALs.

The nice thing about coastlines is that they are difficult to guard. There is no such thing as a straight piece of coast; winds and tides see to that. A 1,000-mile trawl for a ship can be double or triple that distance for a force of soldiers on dry land. The covert entry team need only select a piece that is unguarded and then get ashore. It's not as easy as it sounds, though—it's dangerous work. The submarine noses as close to the beach as it can. The first thing above the water is the search periscope with an ESM receiver, sniffing for electronic signals—radar first of all, then radio communications. If these are identified, the submarine skipper gets moving to avoid both.

The SEALs—the Navy's elite and exclusive SEa-Air-Land commando teams—will probably exit the submarine from underwater using one of the escape trunks. As the SEALs are approaching land with the utmost caution, the submarine captain tries to find a convenient place to wait, perhaps hugging the bottom, probably poking a radio mast up at preset intervals, waiting to recover the returning SEALs when their mission is done.

When the SEALs have completed this mission it's time to return to the sub. Despite what the movies would have you believe, usually the egress phase is quite calm and goes according to plan. If they have committed violence, there will be confusion. If all they have done is to look around and take pictures, then the victims will probably never know they have been had. Once the SEALs are on board, the submarine's skipper quietly leaves the area. Another special operation has been completed, and the joint SEAL/submarine team within the Navy has grown just a little bit closer. And each group of men feels both kinship and distant admiration for the other: the submariners because they have no desire whatsoever to go onto the beach—if they wanted to be Marines, they would have asked for it. The SEALs, on the other hand, shudder at the thought of being inside a steel pipe for weeks at a stretch. It takes all kinds to do the job.

Tactical Example—Special Information Gathering

It was called Ivy Bells. Once upon a time the U.S. Navy learned, never mind how, that there was a telephone cable on the floor of the Sea of Okhotsk that ran from Vladivostok to Petropavlovsk. Both cities were the sites of major Soviet naval bases, and someone, never mind who, wondered if it might be worthwhile to tap that telephone line. And so, an American SSN entered the area.

The Russians claim the Sea of Okhotsk as territorial waters. The United States does not recognize that claim. It's a fine legal point, over what closure rule you think is appropriate. In either case, it's relatively shallow water, a little too shallow for a submarine commander to be completely comfortable, all the more so since the Russians regard it as home waters, hold exercises there, and probably have it thoroughly wired for sound.

But at some time in the late 1960s or early 1970s, a U.S. SSN (perhaps USS *Skate*) made a call and located that phone line. Swimmers went out the escape trunk and made the tap. Then they attached a recording device, probably using an extremely long cassette tape. For the next several years, perhaps extending into the 1980s, a submarine periodically (every month or so) had to reenter the Sea of Okhotsk to download the data on the tape cassette for "processing."

Sure enough, the phone line was used by the Soviet Navy, and so secure did they believe the phone line to be that the data on the line was not encrypted. Everything the Russians knew and did at sea came across that telephone line, and after a brief handling delay, all of that data reached the U.S. Navy's intelligence headquarters at Suitland, Maryland, just outside Washington, D.C., not far from the Smithsonian Institution's Silver Hill Annex.

Of all the intelligence operations conducted by the United States since World War II—at least, all those that have come to the light of day—this is probably one of the most productive, and certainly the most elegant. Which is not to say it was easy. On at least one occasion when a U.S. sub was trying to retrieve the data on the cassette, a Soviet live-fire exercise was underway overhead, and the American crew had no option other than to hug the bottom and hope the Soviet weapons were working properly, because to move away would have presented their counterparts with a target upon which they might have fired live weapons. It became dicier still later. A spy by the name of Ronald Pelton, an employee of the National Security Agency, revealed Ivy Bells to the KGB—for which he was paid the princely fee of perhaps $15,000; the KGB was never generous to its spies—and the tap was discovered. At this writing, Mr. Pelton lives in the basement level of the United States Penitentiary at Marion, Illinois.

What happened when the next submarine went in to download the monthly "take"? That's an untold part of the story. Suffice it to say that Mr. Pelton, in addition to denying his country a hugely valuable source of information, placed over a hundred men at the gravest risk. Was the data worth the risk? Yes. Can submarines still do things like that? What do you think?

Mission #4—Precision Strike: Tomahawk Attacks

As was shown in Desert Storm, a submarine can do many things. Let's say there is a building you don't like. The other guy has lots of radar around it, and maybe the F-117A stealth fighters can't get there. (One needs to remember that the so-called black jet is invisible on radar, but the aerial tankers it refuels from are not.) And you want to do this job with minimum notice to the other side.

A submarine approaches the coast—not all that close, actually—probably at night, and launches a UGM-109 Tomahawk Land Attack Missile (TLAM). For the first few seconds of flight the bird rises rapidly on its rocket booster. Then the wings and tail deploy, the intake for the turbofan engine opens, and the Tomahawk settles down, easily to within a hundred feet or so of the surface. It's a small missile, difficult to detect, especially with the new stealth features added to the Block III missiles now in production. The missile, knowing exactly where it launched from because of GPS satellite fixes (another Block III innovation), then follows a path defined by its inherently accurate terrain-following navigation systems. How accurate will it be? On a good day, a Tomahawk can fly into the door of a two-car garage at a distance of several hundred miles. And that can ruin your whole day.

Tactical Example—Execution of a TLAM-C Strike on an Enemy Airfield

It is not often remembered that the majority of attack aircraft employed by the Imperial Japanese Navy in the Pearl Harbor attack were tasked to counter air missions so that the remainder could attack the U.S. Navy in relative peace. Enemy aircraft are always the most enticing of targets, especially when they are sitting still. But your aircraft also have flight crews, and their lives are precious. That makes them targets also. I will, for once, blow my own horn. I was the first, I think, to consider this possibility in the open media when I included it (as Operation Doolittle) in my second novel, *Red Storm Rising*. (A more professional version was run in *The Submarine Review*, with my permission.) I'd decided that I wanted to do something that was seemingly outrageous but well within the realm of technical capability. So, why not use submarines launching cruise missiles to take out aircraft? This was, according to reports, a mission the Navy lobbied for in Desert Storm, but which the Air Force denied. Thus a few RAF Tornado aircraft were probably lost as a result of the fact that even the USAF wasn't fully aware of what Tomahawk could do.

The only hard part of the operation is timing. You want all the missiles to arrive within a very short time of one another. The accuracy of Tomahawk means that it can fly right down the center not just of runways but also of taxiways, sprinkling cluster munitions (in the case of the TLAM-D version) as it goes, to attack the world's most delicate artifacts—high-performance aircraft. The truly adventurous can aim the TLAM-C versions (with 1,000-lb high-explosive warheads) right at the doors of aircraft shelters. But if you've planned this right, those doors will be open anyway, and many of the aircraft will be in the open, because the whole idea of this sort of mission is to catch the other fellow unaware. There have even been reports of "special warhead" variants of the Tomahawk, including one that fires rocket-propelled conducting filaments over high-tension power lines to short out an enemy's power grid. You see, the U.S. Navy learned its lesson at Pearl Harbor. It's better to give than to receive.

So who might have aircraft that we might not like? Well, consider those perennial western favorites, the Iranians. Since the end of Desert Storm (with its unexpected windfall of Iraqi warplanes), the Iranians have been conducting a truly huge arms buildup. One report even has them trying to buy a regiment of ex-Soviet Backfire bombers complete with heavy antiship missiles. More mundane, but probably a bit more useful (and affordable) are the large number of Su-24 Fencer strike aircraft they have acquired from Iraqi defectors and the Russians. These medium bombers have excellent range and radar, and can be equipped with a variety of air-to-ground ordnance and antishipping missiles such as the Kh-35 (roughly equivalent to the U.S. Harpoon missile). And considering that the Russians will sell almost anything for hard currency these days, you can bet that the Iranians can buy even the latest in CIS missile technology at bargain prices.

Just suppose that the Iranians, having initiated one of their periodic misunderstandings with their Persian Gulf neighbors, begin to hint that they might initiate another tanker war the way they did in the 1980s. And let's just suppose that the Iranians follow habit and decide to hold a live-fire demonstration for television of their newfound capability. They seem to believe that such demonstrations will cause others to bow to their will. More likely, though, it will result in the signing of a presidential finding authorizing the use of force to preemptively remove the Iranian Su-24 threat to shipping in the region.

The question now is just what kind of force to use. A carrier airstrike, long a favorite of presidents, risks the possible loss of aircraft and the death and/or capture of the aircrews. Use of F-117As, so successful and invulnerable during Desert Storm, requires the cooperation of a friendly government in the region to provide basing. And use of long-range B-2As flying directly from a U.S. base, such as Diego Garcia, would place at risk the crown jewels of the Air Force's Air Combat Command. All for taking out a couple dozen fighter-bombers whose net worth would not pay for a single lost B-2A. Surface vessels could launch a TLAM strike but would be sitting there visible after the strike. Clearly what is needed is something discreet and safe for the American attackers. That something may well be a submarine-launched TLAM strike.

To render the airfield useless and destroy the Su-24 Fencers and Kh-35s will probably take between twenty-four and thirty-six TLAMs. Thus a pair of VLS-equipped Los Angeles–class boats will be needed to deliver the missiles. If submarines with the necessary numbers and types of missiles are not already in place, the missiles can be delivered to the boats at a forward base or tender. In addition to the missiles, the submarines will take delivery of the computerized mission plans developed at one of the Theater Mission Planning Centers (TMPCs). This plan, which can be used as is or updated via a satellite link, will have been designed to put the maximum number of TLAMs over the target airfield in the shortest duration possible. It should be noted that not one TLAM will be aimed at the runways. This is because, as Desert Storm proved, it makes little sense to attack concrete, which is quite easy to repair. Destroy an airplane, it is gone forever. And that is the goal of the planned strike.

The run-in to the target probably can be on the coast of the Indian Ocean, though USS *Topeka* (SSN-754) recently operated inside the Persian Gulf itself. The 688Is stand off the coast at a range of 50 to 100 miles and await the firing orders from Washington. Once these come, the firing times and time-on-target of the missiles will be coordinated between the two boats. The mission can be run at almost any time of the day or night, as long as the visibility over the target is relatively clear. For our purposes, though, we can assume that the attack will be mounted in the early morning hours, prior to sunrise. This will have the effect of catching the personnel at the airbase in their beds, reducing collateral casualties as well as the effectiveness of the base defenses.

Each submarine probably loads three torpedo tubes with TLAMs, and only one tube is loaded with an Mk 48 ADCAP "just in case." The three missiles in the tubes will be fired first, followed by the twelve in the VLS tubes. Approximately every 30 seconds another TLAM is ejected from its firing tube and headed on its way. While this is being done, each boat's torpedo room crew quickly reloads the empty tubes with three additional TLAMs, so these can also be launched on their way to the target. This makes a total of thirty-six TLAMs headed for the target airfield. Once this is done, the submarines just slip quietly away, leaving no sign of ever having been there.

Once its engine has ignited and the wings have deployed, each missile skims the ocean and maneuvers to what is known as the pre-landfall waypoint. This a spot in the ocean that leads to the first landfall navigation point. From here, each missile navigates via a combination of GPS fixes and Tercom updates. The idea is for all thirty-six missiles to arrive over the target at precisely the right time and in order. The first few missiles, say four to six of the 1,000-lb high-explosive (HE) warhead TLAMs, are dedicated to reducing the radar and SAM defenses of the airfield. The missiles have been programmed either to dive into a radar and explode, or to fly over the radar vans and destroy them by overpressures created by the high-explosive warheads.

With the way now clear for the remaining missiles, the actual attack on the airfield develops. It will probably be over in just a matter of several minutes. Several of the TLAMs armed with the CEM submunitions will run down the ready ramps, scattering bomblets over any aircraft waiting there. Once each of the missiles has expended its load of submunitions, it will probably be programmed to dive into one of the smaller buildings (such as the aircrew quarters) on the airfield, adding its remaining fuel to the destruction. In addition, each of the large hangars has probably been allocated a pair of HE-warhead TLAMs to destroy any aircraft being serviced there. The fuel storage areas and weapons bunkers also receive the attention of their own TLAMs. The last item for the TLAMs is any revetments or hardened aircraft shelters (HAS) that might possibly contain some of the Su-24s.

Before the base personnel have even had a chance to react, the attack will be over. Most if not all the offending fighter-bombers will be either destroyed or severely damaged. In addition, the antishipping missiles are probably blowing up in their bunkers, and the jet fuel will be blazing in its tanks. And with this, the threat of these aircraft and missiles preying on the tanker traffic in the Persian Gulf will be at an end. All of this has been accomplished without a single American life being placed in harm's way.

Mission #5—Intelligence Gathering

Nobody really listens at keyholes anymore, mainly because keys are smaller than they used to be. But electronics have made the doors rather wide, and they also allow things to leak out more readily than before. The majority of the world's major cities are near the water—they started off as ports and trading centers—and thus are within the reach of submarines and their sensors. Those sensors and their associated analysis equipment help give the United States and her allies an edge in figuring out the policies of foreign governments, and their potential to cause mischief in the New World Order.

Tactical Example—Reconnoitering an Enemy Harbor

It helps to be invisible. That means you can get in close, and when you do that, you can learn things. The prime intelligence-gathering mission for a submarine is electronic surveillance. A simple-looking reedlike mast can gather all manner of electronic signals. You might want to learn about the other guy's radar systems, and he'll be careful with these so as not to let you know exactly what your aircraft will be up against. Therefore he won't use them much when unknown aircraft are about—but he has to use them some of the time in order that his own people can practice using them. And so what you do is sneak a boat into his coastal operations zone, run up your ESM mast, and wait. You can also listen in to short-range radio traffic, the FM stuff that stops at the horizon. Such radios are normally not encrypted, and it's amazing what people will say when they don't think anyone is listening.

In short order, you can monitor the other fellow's whole electronic spectrum, and over a period of time, to boot. This allows operating patterns and procedures to be explored. And you can learn a lot from that. You can do combined operations, with submarines and aircraft working together to see what is really on the other fellow's mind, and you can get away with it because he can see only one element of the operation. Or you can try something really crazy—take a close look for yourself. What is he up to inside his main naval bases? If the water's deep enough, if the sub is quiet enough, you might be able to go in and snap a few pictures through the periscope. Maybe even a few hull shots. Do SSNs ever really do this kind of thing? It's much too dangerous, isn't it?

Mission #6—Mine Warfare

Question: How many mines does it take to make a minefield? Answer: None; you only need a press release. General Norman Schwarzkopf said it all during Desert Storm with one question to an obtuse reporter: "Have you ever been in a minefield?" Imagine what it's like. Every step you take might place you on the trigger of an explosive device. Every single step. You have to get where you want to go. But the simple act of going there may kill you. You don't know when you're entering the minefield, and you probably won't know when you're finally out of it. Sound like fun?

And so it is for ships. A ship, remember, is a steel bubble designed to keep air in and water out. And any ship can be a minesweeper. Once. Mines can be large or small, but in either case they blast holes in ships. Improving technology has made them more deadly. No longer the spherical steel containers with acid-filled horns (though these still exist and still work), modern mines can lie on the bottom, be activated weeks after being laid, and can include special triggering devices so that one might go off when the first ship passes over, and its neighbor when the eleventh does.

Mines, therefore, have a severe psychological impact, and in the natural dread of such things comes panic, concern, and an inordinate degree of effort to get rid of the damned things, a task both time-consuming and very, very iffy. How do you know when you have swept them all? You don't. You can't.

Tactical Example—Quarantining (Mining) an Enemy Port

It only takes a press release, but a single explosion will put a little emphasis on it. Mines are relatively small and compact, and a submarine can carry a goodly number of them, trading off roughly one torpedo for every two mines. And the submarine can deliver a wide variety of them: Mark 57 moored mines with sophisticated sensor and triggering systems. Then there are the Mk 67 mobile mines. These are obsolete Mk 37 torpedoes that have been rebuilt into bottom mines. A submarine can fire them into a shallow channel (which itself might be mined) up to a distance of 5 to 7 miles. The Mk 67 then lies on the bottom, waiting for a ship to pass over it before detonating. Finally, for real impact, there are the Mk 60 Captor mines. These are encapsulated Mark 46 torpedoes programmed to wait for the right kind of noise (in this case enemy submarines), at which point the torpedo swims clear and attacks. For example, you could program them to listen for a certain type of submarine (like a Kilo), which isn't exactly cricket. Mines that shoot first? Just the things for closing a port down.

Let us say that there is a country, North Korea for example, with a nasty habit of exporting military hardware, which offends the sensibilities of the rest of the world. Let's say their nuclear weapons program has finally yielded results. Being strapped for capital, perhaps they might choose to sell off a few to the highest bidder. Somehow (perhaps through some of their contacts in the Swiss banking industry), the American intelligence services get word of the transaction. This starts the ball rolling on a confrontation between the United States with her allies, and the North Koreans. It's the kind of confrontation that the United States could go to the UN with, and make a point to the world about arms proliferation, or suffer a major foreign policy debacle. Not so long ago, the United States expended huge resources tracking a ship loaded with a cargo of North Korean–manufactured missiles on its way to Iran. At the last minute the CENTCOM maritime surveillance forces lost track of the ship, and the cargo was delivered despite the protests of the rest of the world. Would it not have been more effective to just bottle up the port in North Korea with mines and never let the ship out in the first place? That way, wouldn't the UN have a chance to inspect the cargo and make sure it did not contain the offending weapons? You bet! It is a "must win" kind of situation that requires a delicate and discreet touch.

So how does one deploy the mines to close the port in question? The problem here is that the North Koreans have a proven track record of hostility toward U.S. surface vessels and aircraft operating anywhere near their borders. (Remember the capture of the USS *Pueblo* and the EC-121 shootdown in 1968?) Thus it is imperative that any such action be handled carefully. Just the kind of job submarines are best suited for.

The mines are quietly delivered to a 688I at a tender at Guam or some other forward base. The 688I probably offloads all her missiles (except perhaps for Tomahawk antiship missiles in the VLS tubes), and most of her Mk 48 torpedoes. Other than the mines, her only weapons are likely for self-defense. In addition, a SEAL team might be embarked to assist in any on-the-spot surveys required to support the mission. The mining plan has probably been carefully worked out, with appropriate consideration given to such things as the activation times, tidal and seabed conditions, types of mines, and appropriate warnings to the other interested parties involved. Of critical importance is knowledge of the exact placement of each mine, as we would probably have to sweep them (as we did in North Vietnam in 1973) after the incident is closed.

The operation begins with the 688I reconnoitering the areas surrounding the port. Part of this is to establish the operating patterns of North Korean patrols, but also to check for irregularities in the charts and seabed surveys that might affect the mining plan. Here, the Navstar GPS system is critical, as it allows for precise navigation of the boat in the confines of the North Korean coastal waters, and placement of the mines. Once the survey is finished, the job of mine deployment begins.

First out of the tubes probably are the Mk 57 moored mines, to be placed in the outer mouth of the port. The 688I goes in slowly using every sensor of the BSY-1 system to look for trouble. Every few minutes, another mine package is ejected from her torpedo tubes, their activation clocks ticking away to a prearranged time (probably one to two days later). As each mine is released, its position is carefully noted for future sweeping. It will not take many of these, as ship captains are creatures of habit who follow their charts and rarely deviate into less traveled channels. Once this is done, the submarine's commander may fire some of the Mk 67 mobile mines up into the shallow channel leading to the inner harbor, say six to eight of these for each side of the channel, to sit on the bottom. Now the 688I carefully moves out of the area. Just to keep things fair in the coming crisis, the boat might move to one of the nearby naval bases that handles their fleet of diesel submarines and patrol boats. Here it could lay a few more Mk 67s in the channel, and possibly a belt of Mk 60 Captors to keep the North Korean Navy, particularly their force of diesel submarines, bottled up during the coming confrontation. You don't even have to do it to all of their bases. Just do it to one, and say that you have done it to all of them. Who is to know, right?

You now have a foreign policy fait accompli. And don't forget the press release. . . .

Mission #7—Submarine Rescue

It is an acknowledged fact that duty on submarines is more hazardous than other forms of military service. And unfortunately, these extra hazards can translate into the loss of a submarine and its crew. This is the part of submarine duty that is almost never spoken of, even between members of the sub force and their families: if a boat is posted as missing and presumed lost, it probably has been lost with all hands at sea. This was certainly true of submarine losses during the world wars, when *very* few individuals survived submarine sinkings. And in both of the nuclear submarine losses suffered by the United States during the Cold War (the *Thresher* and the *Scorpion*), this precedent held true, with all hands being lost.

Nevertheless, history also tells us that sometimes men do survive submarine sinkings. When the submarine USS *Squalus* sank because of a faulty induction valve off the New England coast in the 1930s, prompt action by the rescue forces of the U.S. Navy saved about half her crew. And when USS *Tang* was sunk by a circular-running torpedo in

1944, a small number of her crew were able to escape and survive until being picked up and taken prisoner by the Japanese. The point here is that circumstances sometimes do allow the crew of a damaged or sunken submarine to survive. And if a navy failed to provide those survivors a chance to live and be rescued, morale in that force would plummet.

So those navies that operate large forces of submarines have invested considerable funds into providing their submariners with equipment and skills to allow for their rescue if they survive whatever initial calamity befalls them. Some of these, like the Steinke hoods and Mark 8 survival suits issued by the U.S. Navy and Royal Navy, are designed for use by the men themselves. But certainly the most visible signs of commitment to the mission of submarine rescue are the Deep-Submergence Rescue Vehicles (DSRVs) operated and maintained by the United States and England. In the wake of the loss of *Thresher* in 1960, the United States built two of these miniature submarines, and the United Kingdom, one. These small submarines, operated from a mother ship or another submarine, can offload the crew from a sunken or damaged submarine and return them to safety.

Example—Rescue of a Downed Submarine

It's a funny thing: most submariners feel that the time of their greatest hazard is during the transits to and from their home bases. This is because of the simple fact that submarines are, by design, hard to see and find. This is especially true when the boats are on the surface, in the transit lanes leading into and out of their lairs. Their low silhouettes and relatively low radar signature make them tough to see. And if a merchant ship crew becomes sloppy or lax, it is quite easy to get run over. The British lost a boat in the Thames estuary in the 1950s, and the French a large cruiser submarine in World War II, to just such accidents. And with the sloppy handling of supertankers that has been so evident over the last few years, it is not hard to imagine an event like this taking place.

So let us suppose that the worst comes to pass, and a merchant vessel, running in heavy fog, collides with a British nuclear attack submarine during a transit back into base at Plymouth. We will suppose that the hit occurs while the sub is running on the surface, striking the after portion of the boat, rupturing the after ballast tanks and destroying the propulsion train. The boat will probably begin to settle from the stern, and there is a good chance of flooding back in the engineering spaces through tears in the hull and the shaft packing seals. With the inrush of water aft, the boat will be headed down to the bottom. During this time, the crew are trying to secure the flooding and seal hatches. The automatic safety systems will "scram" the reactor, making it safe. If there is time, the captain will order the radio room to get off a distress call to the operations center at Plymouth. If not, the crew deploys a buoy, which will transmit its own distress signal to attract attention.

Because of the long continental shelf around the British Isles, there is a good chance that the damaged boat will bottom out in water something less than 1,000 feet deep. Since this is less than the rated crush depth of a British SSN, there is a good chance that some or all of the crew will escape any flooded compartments. At this point, their goal is to survive and wait for rescue if possible. If there is continued flooding, the crew will move to the forward escape trunk, don their Mk 8 escape suits, and free-ascend to the surface. But if the surviving compartments are dry, they will probably try to stay put, hoping for rescue by forces from Plymouth.

Once the Plymouth Operations Center gets the word that something has gone wrong, they set in motion a series of preplanned activities to rescue the downed sub's survivors. One of the first is a call to the U.S. Navy to get the loan of one of the DSRV rescue submarines from SUBDEVGRU 1 at Ballast Point in San Diego, California. As quickly as it can be arranged, a C-5 Galaxy or C-141 Starlifter will arrive at NAS North Island to pick up the DSRV, its crew, and the necessary fittings and equipment to conduct the operation. The idea is that SUBDEVGRU 1 can deliver a DSRV to any point on earth within twenty-four hours, and rescue any crew within forty-eight hours. In this case, the delivery point will be the point closest to one of the "R" class SSBNs, which are equipped to carry and operate the U.S. DSRVs for the Royal Navy. When the transport aircraft arrives, the DSRV and her support equipment are trucked to the port, to be loaded onto a special rack on the back of the British SSBN.

While all this is going on, the crew of the downed submarine are doing their best to do absolutely nothing but stay alive. To purify the air in the surviving compartments, the captain will order the lighting of special candles which, when they burn, release oxygen. Everyone will be ordered to stay quiet, sleep if possible, and just wait calmly. By this time, the Royal Navy has probably assembled a rescue force, which will try to make contact with the survivors and help organize the rescue effort. The first vessel at the site of the sinking may well be another submarine, because of their rapid mobility and their ability to stay on station, whatever the weather and sea conditions. (When the USS *Squalus* was lost in the 1930s, it was a sister boat, the USS

Sculpin, that made first contact with survivors of that downed boat.)

With luck, the "R" class SSBN will able to reach the sinking site near Plymouth within twenty-four to thirty-six hours of the sinking. And at this point, things begin to happen rather quickly. Once the site of the sinking has been established and the attitude of the sunken sub ascertained, the SSBN will submerge and loiter near the downed boat. The crew of the DSRV will enter their boat via the after escape trunk of the SSBN, seal their bottom hatch, and lift off. Since the after part of the sunken sub is flooded, all the survivors will have to exit through the forward escape trunk, and the captain will have to organize the survivors into groups of twenty-four, the maximum the DSRVs are capable of carrying on one trip. At this point the operation begins to look more like two spacecraft docking in orbit. The DSRV maneuvers over the hatch of the downed sub's forward escape trunk and carefully maneuvers down to dock. Once secure, the DSRV's crew blow the water out of the docking collar and bang on the hatch of the escape trunk to tell the crew of the sub that it is time to start the transfer. If the survivors require any medical attention, the DSRV will probably transfer a medical team for the injured. At this point, the first load of survivors enter the two spheres of the DSRV, seal the hatches, and lift off to return to the SSBN. Once there, the DSRV docks with the boomer and discharges the first load of survivors, then repeats the process as many times as required. If the entire crew of the downed sub has survived, it will take four to five trips to offload them all. At this point, any survivors who are seriously injured are MedEvacked via helicopter to a shore hospital.

With the successful rescue of the downed sub's crew, the next job will be to begin salvage of the downed boat.

And have no doubt that this will be done, both for the obvious political reasons, and hopefully to put her back into service. And before you doubt the possibility of such a thing, remember that after the USS *Squalus* was sunk in the 1930s, she was raised and renamed USS *Sailfish*. She would go on, reborn with a new name and crew, to an outstanding combat record, including the sinking of the first Japanese aircraft carrier by a U.S. submarine. Sometimes from the depths of disaster come the tools of victory.

The End of History: Submarines in the Post–Cold War World

What a difference a decade makes. Since the publication of the first edition of *Submarine,* there have been numerous changes to the submarine forces of the world's navies, especially that of the United States. Perhaps the most obvious of these are the introduction of the Seawolf-class (SSN-21) boats into service and the continuing work on a new submarine—the Virginia (SSN-774) class. There have been other dramatic changes as well, especially in the fields of engineering, sensors, and weapons. These advances have led to breathtaking improvements in the way we design and plan submarines of the future. At the same time, they will have a profound impact on the way the Navy's newest submarines will fight the potential battles of the twenty-first century.

Submarine Operations in the 1990s

The decade of the 1990s opened with American submarines supporting their first shooting war since 1945. Operation

Desert Storm allowed the U.S. submarine force to participate in a major conflict, through the use of BGM-109 Tomahawk land attack cruise missiles. The American boats also provided other valuable services during the 1990–1991 war, such as intelligence gathering, maritime surveillance, and special operations support. This trend continued throughout the decade, despite the radical drawdown in the size of the submarine forces of all nations. In fact, the collapse of the Soviet Union and its navy in the 1990s actually freed up the U.S. submarine fleet to undertake a much broader and more significant set of roles in addition to such dangerous yet essential tasks as keeping track of enemy "boomers" and their escorting attack submarines.

Submarines in the 1990s became significant strike platforms, launching Tomahawk attacks into Iraq, the Balkans, and even the retaliation strikes against the Osama bin Laden terrorist organization. So valuable was the capability of submarine-launched cruise missiles that the United Kingdom bought a supply of Tomahawks for their own boats, firing several dozen at Serbian targets during Operation Allied Force in 1999. This covert precision-strike capability has become so attractive that the Royal Navy has looked at equipping *every* British submarine, including strategic ballistic missile boats, with a supply of American cruise missiles. The U.S. Navy has also considered adding Tomahawks to strategic missile boats, proposing to convert the four oldest Ohio-class SSBNs into huge guided missile/special operations platforms.[1]

---

[1] The four boats being considered for conversion to guided missile submarines (SSGNs) include the *Ohio* (SSBN-726), *Michigan* (SSBN-727), *Florida* (SSBN-728), and *Georgia* (SSBN-729), all of which were scheduled for decommissioning under the START-2 arms-control

Another role that submarines have made their own has been in the arena of special warfare and operations. While the British have always used their fleet of boats to deliver and extract special-operations force (SOF) units like the Royal Marines and their Special Boat Squadrons, American nuclear boats spent most of the Cold War chasing Soviet subs and ships. The exception, of course, was the handful of so-called Special Projects boats, which were converted from existing SSNs.[2] However, the end of the East-West conflict and the emergence of the U.S. Special Operations Command as a result of the 1980s defense reorganization acts has changed all that. Today, SOF units from not only the Navy (the famous SEAL teams) but from the Army Special Forces (the "Green Berets") and Marines now regularly practice their trade from nuclear submarines.[3] Two older strategic ballistic missile boats have even

---

agreement. While several configurations are being considered, the basic idea is to fit the Trident missile tubes with seven-cell vertical launchers and storage for supplies to support special operations forces (SOF). Between 126 and 154 Tomahawks would be carried, along with up to 66 SOF personnel.

[2] While a number of boats were configured for electronic eavesdropping, at least four American SSNs were converted into covert operations platforms for deep ocean search and recovery as well as inshore tapping of undersea communications cables. Of the four, only *Parche* (SSN-683) is still in commission and will be replaced in the next few years. The other three, *Seawolf* (SSN-575), *Halibut* (SSGN-587), and *Richard B. Russell* (SSN-687) were all decommissioned prior to, or at the end of, the Cold War.

[3] For more on special warfare and the Army Special Forces, see my book *Special Forces: A Guided Tour of U.S. Army Special Operations* (Berkley Books, 2001).

been converted into transport submarines to support the SOF mission.[4]

Another "growth" mission for submarines in the 1990s has been intelligence gathering, though obviously with less of a focus on the former Soviet Union. The end of the USSR in 1991 freed up American and British boats to keep an eye on a number of other "hot" spots around the world and provide the intelligence services with even more tools and resources to keep an eye and ear on things. One recent example of this probably occurred following the in-flight collision of a U.S. Navy EP-3E *Aries* electronic surveillance aircraft and a Chinese J-8 interceptor over the South China Sea. Though there was a "gap" in the coverage for the U.S. from the air prior to the flights being resumed, rest assured that electronic and communications activity along the Chinese coast was probably being monitored by one or more U.S. submarines. Not only did this fulfill our minimum intelligence-collection requirements, but it also maintained a covert discretion that surface ships and aircraft cannot maintain.

Finally, there are the now-mundane but terribly vital jobs that nuclear boats did throughout the Cold War: watching and tracking the ships and submarines of potential enemies and hostile nations around the world. This has meant that in addition to watching the dwindling fleet of Russian ships and submarines, U.S. and British boats have been keeping an eye on those nations who were quietly de-

---

[4] The two Lafayette-class (SSBN-616) ballistic missile boats, *Kamehameha* (SSN-642) and *James K. Polk* (SSN-645), have had their missile tubes converted to storage areas and been fitted with dry-dock hangars for specialized miniature submarines called SEAL Delivery Vehicles (SDVs). Each can carry and support up to 67 SEALs or troops with everything from explosives to rubber boats.

veloping their own fleets in the 1990s. This may sound surprising, given the worldwide drawdown of naval and submarine forces that followed the end of the Cold War. However, a number of countries began to build up their naval forces in the decade just past, and American and British submarines were out there, watching them every important step of the way.

Into the Twenty-first Century: Submarine Forces at the Millennium

There has been good news and bad news for the submarine forces of the United States and Great Britain. The good news is that due to the demise of the USSR, several regional economic downturns, and the general outbreak of peace, the size of the worldwide submarine force has shrunken to a fraction of its Cold War peak. Literally hundreds of submarines, from ancient diesel boats to state-of-the-art nuclear attack and missile submarines, were taken out of service. In the most radical cases, some of the units from the former Soviet fleet were just driven up onto shore and ditched like whales beaching themselves to die. It was a pitiful ending for the world's largest submarine force.

The bad news is that the submarines that remain in worldwide use are generally the pick of the litter: the best every nation still operating them can afford to maintain. This means that if a shooting war ever breaks out, the boats and captains facing U.S. and British submarine skippers will likely be very capable enemies indeed. Then there is the matter of those nations that have failed to notice the general outbreak of peace in the 1990s. Countries like China, Iran, and India have been building up their navies, and at the core of these efforts have been the expansion of their submarine

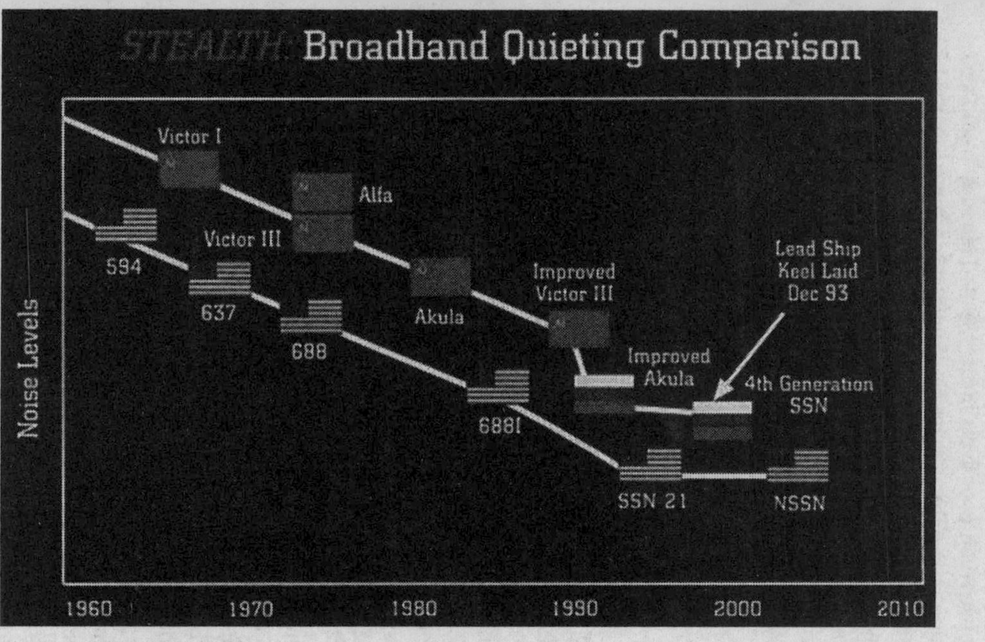

The progressively lower noise levels emitted by various U.S., Soviet, and Russian submarine classes. As can be seen, the Soviet/Russian boats have gotten much closer to the stealth of U.S. boats over the years. *Offical U.S. Navy Graphic*

forces. Many of these have been exported Russian Project 877/Kilo-class diesel/electric submarines (SSKs), armed with some of the best weapons ever offered for sale on the open market.

Similarly, the U.S. and many of our allies are producing the finest submarines, skippers, crews, and weapons that their treasuries can buy. However, these forces will be largely based on "legacy" designs like the 688Is and Trafalgars, with only limited numbers of new boats to replace the many units that were retired in the 1990s. This means that friendly forces will have to make do with what they have while the new designs mature and come into service. While there can be little doubt of the outcome of a one-on-one battle between a U.S. or British SSN and a submarine from some rogue nation, there is always the small chancc that the "bad guys" will score a lucky "kill." The gods of war are a fickle lot, and the worst frequently happens when ordnance begins to fly. Given the public reaction to the loss of eighteen U.S. special operations soldiers in Somalia in 1993, one can only imagine what the national reaction might be to the loss of a billion-dollar-plus nuclear boat and over a hundred sailors. The accidental loss of the Russian *Kursk* (K-141) in the summer of 2000 gave everyone who operates submarines a shock, and something to think about as they headed into the new millennium.

So what does this all mean in terms of numbers of boats? Well, not as many as the leaders of the U.S. or British navies would like, of course. From a Cold War high of almost 100 and 20, respectively, the totals have dropped to around half that. Today, the Americans plan on maintaining a force of around fifty SSNs, while the British are hard-pressed to keep ten to twelve in service. This represents a very small number of platforms to accomplish a large range of missions. We are thankfully without an active naval conflict to fight, and this number will have to do.

The Seawolf (SSN-21) Class: The Ultimate Cold War Attack Boat

Without a doubt, the most advanced submarine ever to enter service did so at a grand commissioning ceremony on July 19, 1997 in Groton, Connecticut. The USS *Seawolf* (given the hull number SSN-21) was to be the touchstone of the U.S. Navy's submarine forces' transition into the twenty-first century.[5] Certainly Seawolf is an impressive foreshadowing of technological advances to come, though this is achieved at an almost unacceptable cost. Also, as impressive as Seawolf is, she is not without her share of disputes and detractors. In fact, Seawolf has often been referred to as the most controversial submarine in American history, and there is a lot of truth to this claim.

Authorized to the defense budget in fiscal year 1989 (FY89), Seawolf was originally intended to be the lead unit of a class of almost thirty boats. This revolutionary sub was designed to succeed the Improved Los Angeles (688I) class attack submarines. As such, she falls into the same class of weapons as the F-22A Raptor fighter and B-2A Spirit bomber: unlimited Cold War designs put into production with little concern for cost at their time of concep-

[5] Technically, *Seawolf* should have been given the hull number SSN-774. However, the Navy's desire to set the class apart as the first of a new century led to the SSN-21 designation. The SSN-774 designation has been now assigned to the lead boat of the new Virginia-class SSNs.

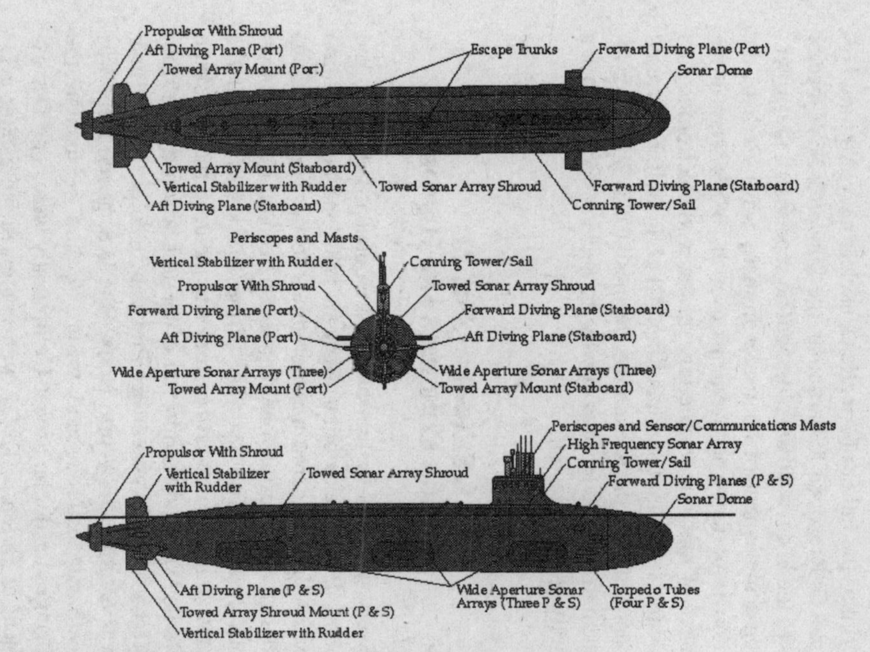

USS *Seawolf* (SSN-21) layout. *Rubicon, Inc., by Laura DeNinno*

tion. In this regard she is a success, as *Seawolf* is reported to be an improvement over the Los Angeles–class boats in nearly every aspect. The biggest pure attack submarine ever built, *Seawolf* also was the last SSN to bear the imprint of the father of America's nuclear navy, Admiral Hyman G. Rickover. In particular, the *Seawolf*'s S6W reactor was the last whose development he supervised, the crowning achievement of his career in many ways. Perhaps the most important improvements over the 688Is were in the areas of machinery quieting, sensors and electronics, and weapons load-out and handling. All of this will be covered later, but first let's examine the post–Cold War environment to get a better idea of why the *Seawolf* became such a hotly debated design.

Seawolf (SSN-21) Design Concepts and History

Every weapons system has a core concept behind it, and *Seawolf* is no exception. Back in 1989, the Soviet Union was still considered a major threat to the United States, though much less of one than it had been during the previous decades. The last year of the 1980s was one of the most dramatic in world history and included the fall of the Berlin Wall and withdrawal of the last Soviet troops from Afghanistan. However, as President George H. W. Bush was entering the White House, the U.S. government was justifiably cautious and unsure of how permanent the changes inside the Soviet Union really were. The Department of Defense (DoD) even continued publishing a famous annual document assessing the Soviet military threat, though with a minor subtitle—*Soviet Military Power: Prospects for Change*—recognizing a possible thawing of the Cold War. The next few years, though, were a very confusing time for the military planners at DoD.

Perhaps the biggest problem DoD faced in the changing global climate was that while the end of the Cold War finally appeared to be a possibility, the American military still needed to prepare for the worst. The Soviet Navy still outnumbered the United States Navy in many ship categories, including the all-important area of submarines. New classes of Soviet submarines seemed to be continually entering service in the 1980s, and the United States simply could not rest on the hope that the Los Angeles–class would forever remain the best boats in the world. It was this environment in which the USS *Seawolf* was conceived.

From a naval combat point of view, it was clearly understood (in fact it was official U.S. Navy policy) that antisubmarine warfare (ASW) was to be the U.S. Navy's *top* war-fighting priority. The Soviet Union and their Warsaw Pact allies could deploy more submarines than the Americans and NATO, and many of their attack and cruise missile submarines were designed with two dangerously important purposes in mind—antishipping and anticarrier operations. The first of these missions dealt with destroying European-bound shipping, along with escorting warships that provided the vital lifeline of the Atlantic—which the United States would have to cross to reinforce and supply its NATO allies in the event of World War III.

The second role of Soviet attack submarines was the so-called anticarrier role, designed to destroy American and NATO aircraft-carrier battle groups, the alliance's most powerful and mobile strike weapon at sea. The Soviets and the Americans both knew that while a carrier's aircraft and escorts might detect and intercept enemy bombers at great distances, the nature of the submarine threat meant that they would not be able to defend against Soviet boats with the same level of efficiency. The role of carrier-hunting submarines, like the missile-armed Project 670/Charlie I/II and Project 949/Oscar I/II SSGNs, was to sneak into range of an American battle group and attack the prey before they were ever detected. This was a very real threat during the Cold War, and it was for this reason that ASW was so vitally important.

One of the best ways to kill a submarine is with another submarine. Continued improvements to the Los Angeles class of submarines were extremely effective and significantly increased the 688I's ability to conduct many types of missions. However, as much as the budget cutters hate to admit it, there comes a time when even the most advanced weapons designs begin to reach their technological limits. In the early 1980s, just as the Navy was ordering the first 688Is, thinking began in earnest about the follow-on to the Los Angeles–class boats.

In the past, much of the silent East-West submarine battle had been fought in the deep ocean depths, far from view of the nearest land. This was to have been the *Seawolf*'s true home. She would be faster, deeper diving, and quieter than any attack submarine the world had ever produced. With ASW already acknowledged as the U.S. Navy's top mission, Seawolf would become the tool for meeting this essential priority. As might be imagined, the project was going to cost some money—lots of money! Initial FY89 cost estimates for the submarine ran in the neighborhood of $39 billion for the full class of thirty boats, which made them the most expensive such vessels in American history. Initial plans called for building three subs per year, which would allow the U.S. Navy to maintain a sufficient force of boats to conduct the required operations in the event the Cold War ever turned hot. It was a good plan, except for the fact that the war it had been designed to fight disappeared within a little over two years.

That type of money—$39 billion—was hard enough to come by during the height of the Cold War and became impossible once it ended. As an uneasy friendship between the U.S. and the former Soviet Union began to grow, so did pressure to trim the American defense budget, which had been slowly declining since the end of the Reagan years. As U.S. politicians clamored for their share of the so-called Peace Dividend, the Defense Department and the Navy began to reexamine exactly what role Seawolf might play in twenty-first-century submarine-force structure. One of the primary lessons learned from the Persian Gulf War was that while submarines were designed to operate in the depths of the blue ocean, there was also an all too frequently ignored requirement for them to support operations on land. Quickly, the Navy began to see the writing on the wall concerning spending on this very expensive weapons program. Several months after the end of the 1991 Gulf War, the chief of naval operations (CNO) announced that Seawolf construction would be cut from the planned three subs a year to a more modest one per year. However, even this plan was modified once the realities of post–Cold War finances and technology began to make themselves known in the 1990s.

The first of the problems for the new class were technical, as might be imagined for such a state-of-the-art weapons system. High-strength HY-80 steel had been used in nearly all previous American nuclear submarine designs since the Skipjack (SSN-585) class of the 1950s. Nevertheless, for the deep-diving *Seawolf*, stronger metal would be needed. Initial plans looked at material as strong as HY-130 steel, but this was eventually shelved in favor of HY-100. The HY-130 was just too hard to work and weld, and production problems with it looked inevitable. Therefore, the Navy and Electric Boat thought HY-100 would be a good compromise between ease of manufacture and greater diving depth. Unfortunately, even the HY-100 steel had its problems when Electric Boat got working. In midsummer 1991, the Navy announced that massive weld failures had been uncovered on *Seawolf*'s hull as it underwent construction. These welding cracks, which might very well have been deadly had they not been discovered and repaired, meant that all welding done to date needed replacement. This caused production of *Seawolf* to be delayed an additional year and added more than $100 million to the already high price of the new boat.

Then further bad news arrived. In 1992, after concluding an agonizing analysis of the situation, the DoD (under then Secretary of Defense Dick Cheney) decided to cut funding for all of the planned SSN-21-class submarines except for the *Seawolf* herself, which was already under construction. As might have been predicted, with the 1992 presidential election looming, the Seawolf program would become a hotly contested political issue.

Running in a tight Democratic primary, a young governor from Arkansas named William Jefferson Clinton announced in 1992 that if elected president, he would save the Seawolf program and continue production past the first unit. Though criticized by some Democrats as supporting a weapons program even the Republicans wanted to cancel, Clinton's gambit paid off. When he won the White House in 1992, he took with him Connecticut's electoral votes, something that might have been impossible without the support of one of that state's biggest employers and its workers—General Dynamics Electric Boat Division—the Seawolf submarine's prime contractor. It is interesting to note that the second submarine of the class was appropriately named USS *Connecticut* (SSN-22).

The year 1992 also marked the beginning of a changing

strategy for U.S. Navy forces. It was during this year that the Navy and Marine Corps released their seminal document that was to serve as a guide for planning the Navy and Marine Corps of the twenty-first century. Entitled *From the Sea: Preparing the Naval Service for the 21st Century,* the document spelled out the biggest change in U.S. naval strategy and policy since the end of the Second World War. Declaring boldly that the Navy's current command of the seas allowed it to concentrate on areas of more likely future conflicts, namely the "littoral" or coastal zones of the earth, the Navy would dramatically alter the planned environment in which they were preparing to fight. In essence, *From the Sea* declared that the ability of the Navy and Marine Corps team to project power from the water and impact events on land would be of dramatically greater importance to future naval planning—more so even than the deep ocean operations of the Cold War. Gone were the days when the so-called blue-water navy took top priority while brown-water units (riverine, mine-hunting, and amphibious forces, among others) languished as a result of a lack of training, funding, and attention from the senior leadership. Impacting events on land was something the U.S. Marine Corps and SEALs community had done for decades, but it was something the majority of U.S. Navy officers and sailors had to learn quickly if the Navy was to have a seat at the table when new conflicts erupted.

While this confident new plan was essential to the Navy's future, it was not good news for the Seawolf program. Seawolf had been designed to fight in the ocean depths and to hunt Soviet submarines. To this end, it was the quietest, deepest-diving attack submarine America had ever planned. The problems facing warships in the shallow, murky "brown" water of the coastal regions were entirely different from those encountered in the open ocean. This was especially true for submarines, which relied on deep diving depths and sensitive passive sonars to maintain their stealth—both of which would be of limited use in the brown-water combat environment.

As if this was not bad enough, the new Seawolf-class boats were also extremely expensive. Practically every part of the Seawolf's design was controversial. While this was largely due to the high cost of incorporating advanced systems into a revolutionary design, there seemed to be problems all along the way, which some of the world's best experts were put to work solving. The result was a truly amazing piece of machinery, which has run taxpayers somewhere in the neighborhood of $2.8 billion per unit of the class. While this may sound like a lot, take into consideration that the Air Force paid $2 billion for each of twenty-one B-2A Spirit stealth bombers.

Fortunately, plans for a new submarine that would incorporate the technology of the *Seawolf* into a boat the size and cost of the Los Angeles class were already in the works. Faced with these fiscal and strategic realities, the Navy put new emphasis on a submarine they were calling *Centurion*—today known as the Virginia (SSN-774) class. It thus came as no surprise when, in October 1993, Secretary of Defense Les Aspin released the results of the Bottom-up Review (BUR) in which it was explained that Seawolf production would end after only three boats, holdovers from Clinton's election '92 promise, had been constructed.

Seawolf (SSN-21): A Guided Tour

Once you get over the sticker-price shock (something Congress never seemed to do!), you can discover exactly how

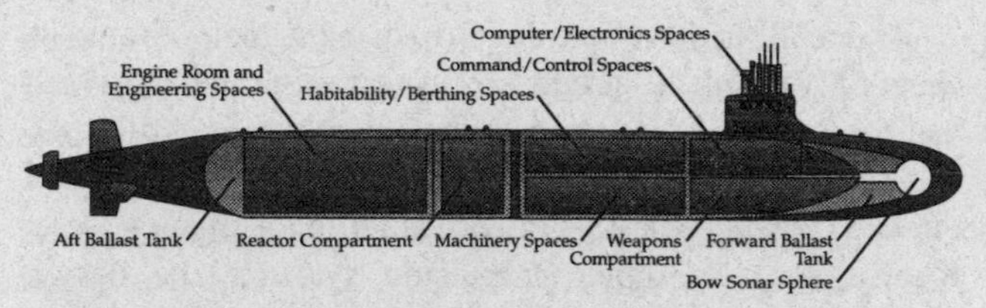

USS *Seawolf* (SSN-21) interior layout. *Rubicon, Inc., by Laura DeNinno*

revolutionary *Seawolf* actually is, and it's something of which every American can be justly proud. Let's start out by discussing the design of this big, beautiful boat. Prior to *Seawolf*'s design, every class of U.S. submarine since the Skipjack (SSN-585) class of the 1950s had been an "iterative" design. That is to say, the basic design of submarines was modified so that each new class was based on the solid design of an older ship, incorporating a mix of old and new technology.

Thus the classes between the Skipjack and Los Angeles were all modified designs of the same original boat. This all changed with the *Seawolf* design. *Seawolf* was the first submarine design in over thirty years to be planned totally new from top to bottom.

Everything about the *Seawolf* (SSN-21) and her sister boats, *Connecticut* (SSN-22) and the not-yet-in-service *Jimmy Carter* (SSN-23), is new and improved. She is big, displacing an impressive 9,137 tons submerged. Starting from the stern, we begin our look with what many people would mistakenly think is one of the simplest parts of a submarine: its screw. Known as a propeller to those outside the Navy, the screw is actually one of the most complicated parts of a submarine, and its construction is a closely guarded national secret. The construction of the *Seawolf*'s screw has been essential to her requirement for quiet running at high speeds.

As mentioned earlier, the British built their Trafalgar-class SSNs with a shroud covering the propeller, which had the benefit of quieting excess noise generated by the sub's screw out into the water. A similar design is used in the U.S. Navy's Mk 48 torpedo, albeit on a smaller scale. Known as a "pumpjet propulsion system," the design works well. According to one report, running at 25 knots, *Seawolf* is quieter than a 688I that is just sitting at the pier! Other stories indicate that *Seawolf* is able to run quietly at twice the speed of any previous American attack submarine. Other sources are more direct and attribute to *Seawolf* a virtually "silent speed" of 20 knots. While numerous elements go into these quieting secrets, you can bet that the *Seawolf*'s pumpjet propulsor plays a key role. Hidden inside the covering shroud of the propulsor is a single propeller shaft similar to those that have been used by U.S. attack submarines since the advent of nuclear propulsion.

If you were to look at a photograph of a Seawolf under construction or in dry dock, you would be able to see many of the sub's sensors as you glance at the sides of its hull. In particular, the boat is designed with a unique surface tail configuration and gives the impression of six thin, flat "fin stabilizers" jutting from the aft of the boat, which face out at varying angles from the shrouded prop. Fitted to the stabilizer at the four- and eight-o'clock positions are shrouds through which the sensitive TB-16D and TB-29 towed array sonars are streamed out from the boat. As you move around to the sides of the lower hull, you'll notice one of the biggest advances perfected between the construction of the last of the 688I boats and the new *Seawolf*. This is the addition of the BQG-5D Wide Aperture Array (WAA) system sensor fittings. Although invisible when the sub lies in the water, the WAA is one of the most distinctive features of this revolutionary warship design. An advanced passive sensor system fitted into three rectangular housings attached to each side of the lower hull, the WAA performs an essential mission when the boat is in the detection and tracking phases of an engagement, and Seawolf is the first full class of submarines fitted with the system. The WAA has been so successful in trials that plans currently call for fitting it into the future Virginia (SSN-774) class as well.

In the bow is a large, 24-foot/ 7.3-meter-diameter spherical sonar array, which is the heart of the BSY-2 combat system. Based on the earlier BSY-1 system we showed you aboard *Miami,* BSY-2 is, in terms of software, processing power, and integration, a generation ahead of the earlier system. By tying together all the various sonar and other sensors systems into the BSY-2, *Seawolf* has a capability for multitarget combat engagements and situational awareness matched only by the Aegis combat system on the *Ticonderoga* (CG-47) and *Arleigh Burke* (DDG-51) missile cruisers and destroyers.

As we continue along our journey on *Seawolf,* you'll notice many bulges and bumps along the hull, each of which serves a vital purpose. Walking along the long hull, which is 353 feet/107.6 meters long and 40 feet/12.2 meters wide, you'll see a long, thin faring that is raised several inches off the deck. This is where the towed array sonars are stored. Also, as during our visit to *Miami* and *Triumph,* you'll notice that the deck is made of a thick, spongy coating known as anechoic tiles. These black rubberlike tiles do much to seal sound inside the *Seawolf* as well as keep other sounds from bouncing off the boat and reflecting sonar "pings" back to prowling surface ships, sonobouys, or enemy submarines. Every now and then you might see a submarine, especially those of the former Soviet Union, missing a tile or two. These occasionally fall off and make for some interesting photo ops!

Underneath the tiles is one of the hardest steel hulls ever constructed on an American ship. Once the welding work of *Seawolf* was fixed, the real benefits of HY-100 steel became apparent. With a significant (meaning *classified*) increase in diving depth over the 688I class, *Seawolf* is able to operate farther into the ocean depths than any attack submarine in American history. This has restored much of the tactical capability lost when the HY-80 hulls of the Los Angeles–class boats were thinned down to save weight and displacement. As the recent loss of the Russian submarine *Kursk* illustrates, the ocean depths can be anything but hospitable, and the deeper a submarine goes the more pressure is exerted on its hull. It was just these dangers that the submarine designers had in mind when they built in the next feature we run across as we tour *Seawolf*'s deck—the submarine escape trunk and Deep Submergence Rescue Vehicle (DSRV) mating hatch.

This aft hatch, along with a second hatch farther forward, is where a rescue chamber or submarine like the DSRV would mate with *Seawolf* in the event she suffered a catastrophic accident and the crewmembers were still safe. This is, of course, a really big if. It is, however, a very real possibility that was demonstrated quite sadly by the loss of the *Kursk* and her crew in 2000. A number of the *Kursk*'s crew survived the sinking of their boat and might have been saved had their government allowed U.S. or British DSRVs to be deployed earlier during the search-and-rescue operations.

There have been some changes in the field of submarine rescue since the first edition of *Submarine* went to press, and this seems like a good place to cover them. The first is that the American DSRVs are rapidly coming to the end of their useful service lives and require replacement or upgrade. Also, the dedicated rescue ships that could operate the old McCann rescue chambers have been retired, meaning that the DSRVs delivered on the backs of submarines are now the only deep-water rescue system in the U.S. inventory. On the plus side, though, new rescue technologies are being designed and tested, and may be backfitted onto existing DSRVs.

One of the most promising of these is a new kind of mating collar, composed of angular slip rings that allow docking even if the downed boat is resting at a severe angle. Whether this new system will be retrofitted to the existing DSRVs as part of a comprehensive overhaul or to a completely new vehicle remains to be seen. For now, though, submarine rescue still remains an "iffy" proposition at best.

Farther forward is the sail, which is, frankly, one of the slickest such structures ever built onto a U.S. submarine. Unlike traditional American nuclear subs, *Seawolf* has a curved faring blending the front of her sail into the hull to help reduce resistance and flow noise. It is just one of many little touches designed to keep the Seawolf-class boats the quietest ever to roam the world's oceans. As in previous American SSNs, the sail contains all of the sensor masts, as well as the control station for conning the boat on the surface. The mast-mounted sensors include:

- **Periscopes:** As in previous U.S. submarine designs, the *Seawolf* is equipped with a pair of optical periscope masts. These include both Type 8 Mod 3 and Type 18 scopes, of the same variety as those described earlier on *Miami*.

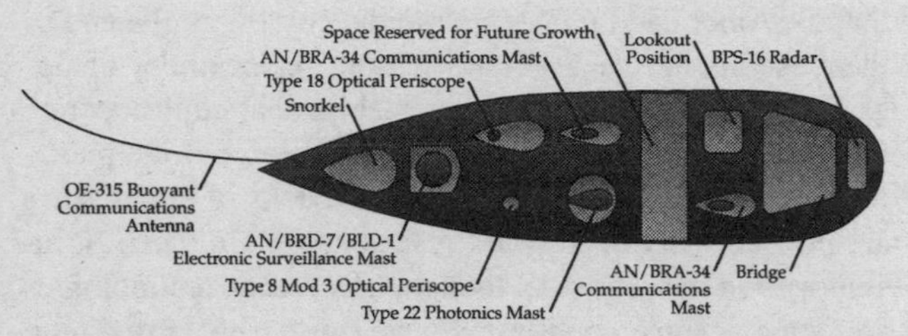

The arrangement of periscopes, sensors, and communications masts on the conning tower/sail of USS *Seawolf* (SSN-21). *Rubicon, Inc., by Laura DeNinno*

- **Radar:** To provide surface and some limited air search capabilities, a BPS-16 set is installed for operations in poor visibility and at night.
- **Radio Masts:** A pair of AN/BRA-34 communications masts are provided to support the growing bandwidth requirements for littoral operations.
- **Electronic/Signals Collection Masts:** To support intelligence collection and tactical situational awareness, *Seawolf* has an AN/BRD-7/BLD-1 mast with the collection heads for the WLQ-4 (V)1 and BLD-1D/F radar and signals receptions systems.
- **Trailing Antenna:** To provide command cueing while submerged, the *Seawolf* has an OE-315 trailing wire antenna that can receive transmission from the Navy's Extremely Low Frequency (ELF) communications system.

All this, along with the improved processing and display technology of the BSY-2 combat system, makes *Seawolf* a truly revolutionary design—and just think, we've not even touched on the weapons load yet! That, too, is a major improvement over that of the older 688Is.

As we continue with our "hull walk," you'll probably notice a large hatch directly aft of the sail structure. This is the oddly shaped weapons shipping hatch and is used in the slow, monotonous process of loading torpedoes, weapons, and other stores inside the boat. One by one, each of the torpedoes (up to a maximum load of fifty) and other weapons must be brought down into the sub and laid in the torpedo room for storage in the event of combat. The weapons load of the *Seawolf*, twice that of the Flight I Los Angeles–class boats, was mandated by the desire to have enough warshots to sustain multiple engagements during prolonged wartime operations. To get these weapons off the boat quickly, *Seawolf* is equipped with eight 26.5-inch/673mm torpedo tubes, the biggest ever fitted to an American submarine. Utilizing a new air turbine pump system to expel the weapons more quietly than earlier water-ram methods, the new tubes are also capable of launching unmanned surveillance vehicles and even divers, should that be necessary. One thing the 688Is had that has been deleted from the *Seawolf* is the bank of Vertical Launch System (VLS) missile launchers in the bow. With her huge internal weapons stowage and eight torpedo tubes, the *Seawolf* was considered well enough armed to eliminate the VLS tubes.

As with every other element of submarine technology, ten years makes a big difference in weapons. Since the early 1990s, there have been significant changes and improvements to the weapons carried by the Seawolf-class boats. First off, all of the UGM-84 sub-Harpoon antiship missiles have now been withdrawn from service in the U.S. submarine fleet. This is mostly due to the fact that each Harpoon takes up space that might be used to hold a more frequently used torpedo or Tomahawk cruise missile, something that has made the sub-launched version of these formidable weapons go the way of the dodo bird.

While not completely making up for the mid-range surface ship attack capability afforded by the Harpoon, the Navy has been hard at work improving their supply of Mk 48 ADCAP torpedoes. These newest modifications to the already advanced torpedo are known as the ADCAP Mods 5 and 6. The Mod 5 changes include a guidance and control modification that improves the acoustic receiver, adds memory to the internal computer, and allows the torpedo to handle increased software demands. The second modification, known as Mod 6, includes the TPU or Torpedo Propulsion Unit upgrade and will provide the ADCAP with greater speed, range, and depth. These improvements to the Mk 48s will enable the weapon to better conduct operations in the coastal zones where the Seawolf-class boats will be lurking and working.

As mentioned earlier, one of the major missions of SSNs in the 1990s has been that of launching BGM-109 Tomahawk cruise missiles against enemy targets. The preferred version, known as Block III, has a GPS-based guidance system as well as a new warhead and satellite telemetry system. The problem is that many of the Tomahawks modified to the Block III standard were fired during the 1990s in places like the Balkans and Southwest Asia, and earlier variants lack the easy mission-planning capabilities of the newer missiles. Several plans were put forth to modify more of the early model missiles to a so-called Block IV configuration, but would have cost too much (over $700,000 per missile).

To provide both surface ships and submarines with enough of the precious Tomahawks into the twenty-first century, a brand-new version, known as Tactical Tomahawk (TACTOM), is being developed by Raytheon. TACTOM will incorporate a number of new features, including a new injection molded plastic airframe, satellite data link, and turbojet engine, to reduce costs. At around $500,000 a copy, the new missiles will be a bargain compared with reworking older airframes. However, the Block IIIs will be the primary variant until the middle of the decade, when TACTOMs should begin to arrive in serious quantities out in the fleet.

If you duck down inside the hatch aft of the sail, at first you will feel just like you have stepped into any other advanced submarine. However, moving forward into the control room, you rapidly can see the differences between Seawolf and the Los Angeles–class boats. Where older boats still have a lot of conventional dials, gauges, and other readouts, most of the critical control positions on *Seawolf* have been equipped with red plasma computer displays with touch screens. These allow a much wider range of controls and graphics to be fed to operators in the control room and other parts of the boat, and stand up quite well to errant elbows and spilled coffee!

Otherwise, the basic layout of *Seawolf* is very similar to that of *Miami,* with perhaps a bit more elbow room than the older boat. Nevertheless, *Seawolf* still does not have all the creature comforts you might expect on a submarine with over 25 percent more internal volume than a 688I. The problem is that while there is more room inside of *Seawolf*, there also is more "stuff" inside her hull. The S6W reactor, while the same basic unit as the one on the 688Is, now feeds two steam turbines putting out an additional 10,000 horsepower. This provides a total of 40,000 shaft horsepower, giving *Seawolf* a top speed of around 35 knots, if you believe reports from the initial sea trials.

These engines in turn have more quieting mounts and equipment than those on the Los Angeles–class boats, all of which take up lots of space. Virtually every other piece of machinery on *Seawolf* has similar quieting gear, which eats up a lot of volume. The result is that a number of the junior enlisted personnel still have to "hot bunk," due to a shortage of berthing space. This is a shortcoming, which will probably have a downside in the long run, in terms of habitability and personnel retention. However, it is the price that must be paid to make *Seawolf* the quietest, most deadly submarine in the world.

The rest of *Seawolf* is much like that of *Miami,* though put together very differently. Electric Boat, the prime *Seawolf* contractor, designed her to be built with a modular construction technique, much like that of Newport News in Virginia and Litton-Ingalls in Mississippi.[6] This means that more of the boat can be "stuffed" and finished before the hull is welded and floated into the water. It would have been interesting to see what this would have done to production costs if even a second flight of three SSN-21s had been ordered by Congress, instead of proceeding to the Virginia (SSN-774) class boats directly. As it is, the sailors assigned to the *Seawolf* and *Connecticut* consider themselves very lucky sailors indeed. Both are in the water and assigned to the Atlantic fleet, starting to make patrols and being tested in exercises.

The *Jimmy Carter*—the third and final Seawolf—will, however, be something very different: a true "Special

---

[6] For more on these two yards and how they build ships, see my books *Marine: A Guided Tour of a Marine Expeditionary Unit* (Berkley Books, 1997) and *Carrier: A Guided Tour of a Carrier Battle Group* (Berkley Books, 1999).

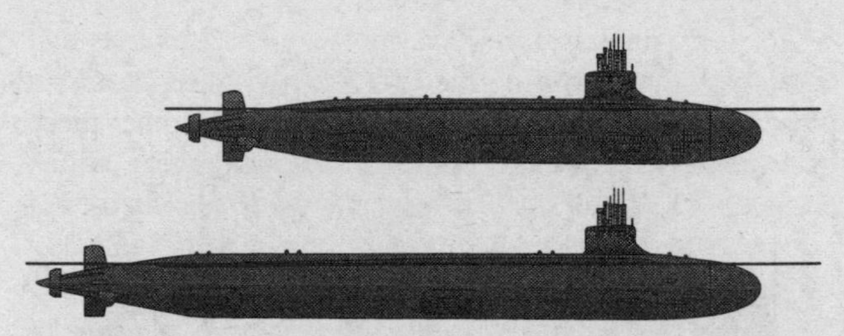

A comparison of the USS *Seawolf* (SSN-21) (*top*) and the USS *Jimmy Carter* (SSN-23) (*bottom*). The roughly 100 foot/30.5 meter greater length of the *Jimmy Carter* will accommodate a "plug" to conduct "Special Projects" and "Research" missions. *Rubicon, Inc., by Laura DeNinno*

Projects" boat from the keel up. The basic *Seawolf* hull is having an approximately 100-foot/30.5-meter "plug" added aft of her sail, with all kinds of room for berthing of extra personnel, stowage of special equipment and sensors, as well as a large lock-out chamber. This will be big enough to allow the launching of the new generation of Unmanned Underwater Vehicles (UUVs) being developed for use by the fleet. The plan is to have her in the water by 2004, when she will join *Parche* (herself scheduled for retirement in 2006) at the Trident-missile submarine base in Bangor, Washington. *Jimmy Carter* will be the ultimate expression of American submarine intelligence gathering, though just what that will mean in the twenty-first century is still unknown. However, given what the Navy's small force of special projects boats did during the Cold War, the *Jimmy Carter* will be doing things that will someday be a subject for novelists.

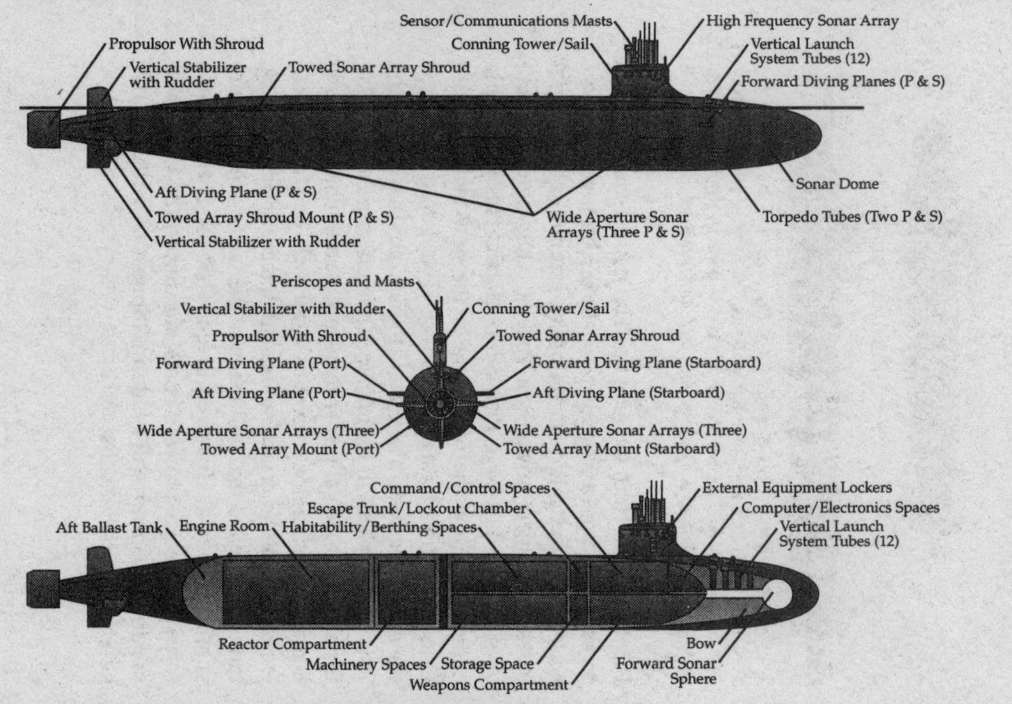

The interior and exterior layouts of the USS *Virginia* (SSN-774). *Rubicon, Inc., by Laura DeNinno*

The Virginia (SSN-774) Class Boats: The New Generation

In the mid-1990s, when only the three Seawolf-class boats were authorized for construction, the Navy realized it clearly had a problem on its hands. How was the submarine service to meet its quantitative requirements for keeping approximately fifty submarines in the fleet? At the end of the Cold War, the U.S. Navy had a goal of 100 SSNs (excluding the strategic missile boat force) as part of a 600-ship Navy. While neither of these goals was ever reached, by the late 1980s the submarine force was very, very close to achieving its force structure goals. In 1987, for example, the U.S. Navy attack submarine force consisted of ninety-nine nuclear attack boats.

All this changed in 1993 when DoD released the results of the Bottom-up Review (BUR), which, attempting to alter the military to a post–Cold War force, cut a little too close to the bone for the comfort of those in the submarine community. Calling for new submarine-force levels as low as forty-five submarines, the BUR drastically changed the goal of the Navy's submarine force away from acquisition and force enhancement toward drastic cuts and getting rid of old boats. The resulting dearth of submarine construction in the mid-1990s meant that only a handful of new boats were finished. However, 2004 promises to be the best year in a long while for the U.S. submarine force. That will be the year the first Virginia-class SSNs enter the Navy and the year the *Jimmy Carter* is commissioned. However, it has taken a very rough dozen years even to see the promise of 2004 for the submarine community.

The curtailing of Seawolf production to just three units meant that the Navy would inevitably have to develop a smaller, more cost-effective design that would better fit the roles and missions set out in *From the Sea*. Fortunately, a series of design studies was already under- way at the time, the most promising known as Centurion. From the beginning, Centurion, whose name was changed several times before she was officially named the Virginia (SSN-774) class, was an easier sell than was the *Seawolf*. The concept behind the Virginia was to build a submarine as good as the *Seawolf* in the bluewater environment, yet able to conduct operations in the littoral regions of the world. Additionally, Virginia needed to counter the biggest drawbacks of Seawolf—its costliness to produce and the fact that it was built by a single yard. The monopoly on construction of the Seawolf class by Electric Boat rankled the folks at Newport News Shipbuilding, and also their powerful congressional delegation.

Costs drove Virginia's design to a far greater degree than any submarine designed for the U.S. Navy. While it possessed roughly the same capabilities and quieting as Seawolf in a more affordable and multimission configuration, initial plans called for the class to be built at an optimistic rate of two or three per year. Original cost projections aimed for a boat displacing roughly 6,000 tons, costing around one-half that of Seawolf.

The hope was—and still is—that this design will do for submarines what the relatively moderately priced, multirole F/A-18 Hornets did for naval aviation. Current plans call for the Virginia class to consist of thirty units, which will be constructed at varying rates for staggered delivery. The first of the class, *Virginia* (SSN-774), will enter service around 2004; the second, named USS *Texas* (SSN-775), will follow a year later. After a one-year break, the current schedule calls for USS *Hawaii* (SSN-776) to join the fleet in 2007, followed by USS *North Carolina* (SSN-777) in 2008. While plans inevitably fall by the wayside and are continuously altered, this seems to be a great start.

At first, the idea was that the Navy would buy the Virginias to complement rather than succeed the Seawolfs. The resulting budget cuts and cost overruns on the SSN-21 program turned out to be so severe, however, that the Navy saw no choice but to move ahead with Virginia after Seawolf production ended. The DoD directed in 1992 that the Navy should hold the costs of the new submarine design to a maximum of $1 billion per boat. The DoD also charged the Navy with examining alternatives to this entirely new class of warship. These started off with a baseline (for comparison purposes) of continued SSN-21 production at a rate of one per year. The alternatives included:

- A lower cost variant of the Seawolf.
- Further improved versions of the Los Angeles (688I) class.
- The possible procurement of non-nuclear (i.e., conventional) submarines into the fleet.

Hyman Rickover must have been turning over in his grave at such thoughts, but then he never lived to see the post–Cold War world of the 1990s! In the end, the Navy stayed committed to the Centurion design, though not without a lot of pressure from critics and Congress.

In 1993, the name Centurion was officially changed to "New Attack Submarine" and given the abbreviation NAS, which was later changed to NSSN (for New SSN). The following year, the Navy began to provide the first real cost estimates on the class. These indicated that the lead submarine, which would be authorized in the FY98 budget, would cost $3.4 billion, including nonrecurring research and design costs. This was as much as *Seawolf* herself, and some people wondered if the Navy should have built more of the SSN-21s instead. However, the Navy study indicated that additional NSSN-class boats, starting with number five, would cost around $1.54 billion in FY98 dollars. While this was still slightly higher than the projected goal, it was far below the $2.8 billion for a production Seawolf, had such a thing ever been built. With the planned production costs now under control, it appeared as if NSSN might actually become a reality.

As with any multibillion-dollar decision, the construction issues of the Virginia class were now beginning to point away from the operational side and toward the financial and political ones. Shipbuilding, especially submarine manufacturing (and particularly that of *nuclear* submarines), is a field that is especially difficult for a nation to master. Several generations of American shipbuilders have been toiling on nuclear submarines since work began on the program under the leadership of Admiral Hyman Rickover. If production were to suddenly end or shrink to less than one boat per year, one of the two American submarine manufacturers, Electric Boat or Newport News Shipbuilding, would inevitably be forced to leave the business. With their powerful political supporters, both contractors began a spirited competition for the right to produce the NSSN boats.

Long ago, Congress had determined that it was in our nation's interest to maintain a minimum of two shipyards capable of building nuclear submarines. This industrial-base-preservation argument was an important key to keeping Seawolf alive for three boats and became equally important to the idea of a "teaming" arrangement between the two American submarine manufacturers. As a result of the 1993 BUR, DoD concluded that it would be ill advised to consolidate all submarine construction at just one shipyard. This was probably a good decision for the Navy because it preserved at least a semblance of competition between Electric Boat and NNS. Additionally, because nuclear submarine design and production is so complicated, it is extremely manpower-intensive. Therefore, any loss of production or a strike at a yard meant inevitable layoffs of highly skilled workers at subcontractors. A shipyard might even be forced to close its doors forever if production levels continued to drop. This has happened dozens of times to some of the biggest shipyards in the nation. One only need remember the demise of great names like Todd and Kaiser on the West Coast to realize that American shipbuilding hangs by a slender thread these days.

How then, was the Navy to keep both submarine yards alive with so few submarines to build? Not surprisingly, the Navy didn't have to look too far for help. It came directly from the two shipbuilders themselves—Electric Boat and Newport News. These two companies knew that it was in the nation's interests—and their own—to solve this dilemma. Which is exactly what they did in December of 1996 when Electric Boat and Newport News Shipbuilding offered the Navy a deal. How would the government like it if the two companies "teamed up" and produced the Virginia class together?

It was a remarkable offer and one that the Navy could not refuse. Both shipbuilders would take advantage of the Digital Design Database Electric Boat had used in constructing the Seawolf class, to help keep down costs and keep quality up. Additionally, each yard would build specific parts or sections of the boats, while each company would build and "stuff" their own reactor plant modules. The bow, stern, and sail sections of all the new subs would be built by NNS, along with the habitability and machinery spaces and the torpedo room. Electric Boat, as the prime contractor, would, in turn, construct nearly all remaining portions, including the engine room and control spaces. Finally, Electric Boat would assemble the first and third boats while NNS would handle the second and fourth.

The current plan calls for teaming on only the first four boats. The Navy, Congress, and the two remaining submarine builders will eventually have to examine where they go from there. Such construction decisions are momentous indeed. Especially when you look at how large a role the Virginias will play in the American submarine force of the twenty-first century.

USS *Virginia* (SSN-774): A Virtual Tour

Now that you have seen the future production plans for this new class of SSN, let's take a look at what new things they will be capable of accomplishing. We have to do this in a virtual fashion, as the Virginias are still mostly "paper" submarines. The official "keel laying" of *Virginia* (SSN-774) only occurred on September 2, 1999, and she will not be delivered to the fleet until 2004. Officially, though, the Pentagon has provided us with a great starting point in its new public campaign of openness about submarine operations and weapons. Gone forever are the days when the "Silent Service" was truly mute to the world outside of their pressure hulls.

As one might have guessed for a multimission boat, the Navy has bestowed upon the Virginias nearly every possible submarine mission under the sun, which speaks volumes about their versatility. These missions, according to the Pentagon, include taking the enemy by complete surprise while conducting:

- **Covert Strike Warfare:** Hitting inland targets with Tactical Tomahawks and possibly future battlefield support missiles.
- **Antisubmarine Warfare (ASW):** Destroying enemy submarines while conducting area underwater surveillance.
- **Covert Intelligence Gathering and Surveillance:** Keeping American eyes on potential hot spots and enemy operations.
- **Antisurface Warfare (ASUW):** Clearing the sealanes of enemy surface ships.
- **Covert Mine Warfare:** Laying, detecting, and possibly clearing friendly and enemy naval minefields.
- **Battle Group/Amphibious Group Support:** Protection and support for Carrier Battle Groups and Amphibious Ready Groups.

- **Special Warfare Support:** A whole variety of clandestine missions, including direct action raids, reconnaissance, combat search and rescue, directing air strikes, and tactical intelligence gathering.

While this is just a brief list of missions *Virginia* can "officially" carry out, imagine the potential for a class such as this. There are even discussions about making the Virginia design the basis for a new class of SSBNs to replace the Ohio-class boats, should this be required.

The general layout of the Virginia-class boats will not be unfamiliar to those who have been aboard previous nuclear attack submarines. In many ways, she is a Los Angeles–sized hull packed with systems pioneered by the Seawolf-class boats. The biggest difference with Virginia is that flexibility is the key in her design. In addition, the use of Commercial Off-The-Shelf (COTS) technology has been maximized in order to reduce overall production costs and allow for rapid integration of new systems and software. Another key element to reducing production costs has been the use of Computer Aided Design (CAD) for this class of submarine. In fact, the *Virginia* will be the first American warship designed solely by computer. In her own way, she is as much of a revolution in construction and systems as *Seawolf* was, with controlled cost management thrown in to tighten everyone's belt a bit!

We'll start our brief look around the *Virginia* at the heart of the boat's power. The new S9G pressurized water reactor produces sufficient shaft horsepower for a top speed only slightly less than that of Seawolf. This is one of the few areas where decreased cost has been allowed to reduce *Virginia*'s capability. The reactor runs two steam turbines geared to a single shaft. In turn, this shaft will connect to a very quiet pumpjet propulsor, similar to those found in the British Trafalgars and the Seawolfs.

Equally importantly from a maintenance point of view, *Virginia*'s reactor will have a "life of the ship" reactor core, meaning there should *never* be a need to replace the reactor core! The new reactor design has been simplified and, amazingly, it should match SSN-21's impressive quieting levels within a 25 percent smaller volume. Because the overall design has been simplified, fewer components are required and the *Virginia* has fewer pumps and valves than any of her predecessors. This smaller size is one of many improvements that allow her overall submerged displacement to remain below 8,000 tons—a 1,000-plus ton reduction from that of the Seawolfs.

The hull of the *Virginia* is made from similar-strength steel as *Seawolf*, although because of *Virginia*'s littoral-operations emphasis, the steel does need to be as thick as *Seawolf*'s—providing additional cost and weight savings. The sub's maneuvering performance will also be unprecedented for a boat her size. The *Virginia*'s control surfaces will be part of a digital "fly-by-wire" ship control capability similar to those used so effectively on fighters such as the F-16 Fighting Falcon and F/A-18 Hornet. This eliminates much of the heavy cabling and hydraulic piping that runs throughout older classes of SSN.

Though *Virginia* is not faster than *Seawolf* and does not dive deeper or even carry as many weapons (only thirty-eight versus fifty for the SSN-21 boats), these facts can be misleading. The real difference between the two subs lies in their ability to conduct operations and fight in the Navy's new home of the littorals. While still able to fight with much of *Seawolf*'s prowess in the open oceans, *Virginia* has incorporated special weapons, sensors, and other new equipment particularly well suited to her coastal missions.

The sonar suite on board the *Virginia* will include special high-frequency acoustic sensors designed to hunt and classify both diesel-electric boats and those with advanced air independent propulsion (AIP) systems. As with other SSNs, *Virginia* will carry a spherical active/ passive sonar array along with the TB-29 Thin-Line Towed Array and the TB-16 Fat-Line Towed Array. *Virginia* will also carry a new lightweight WAA system, specifically optimized to locate super-quiet diesel-electric and AIP submarines. In the littoral regions, special attention will need to be devoted to naval mines, so *Virginia* will have a high-frequency sonar suite to detect the deadly "weapons that wait." The sensors will include sail- and chin- (beneath the sonar sphere) mounted arrays to provide the new boats with their best-ever mine-detection and -avoidance capability. Along with her sonar suite, *Virginia* will also have an array of sail-mounted sensors, almost identical to that of the Seawolf-class boats.

If there is one piece of equipment many submariners thought would never change, it was the old-fashioned periscope. Well, the Navy has got news for us periscope lovers—even this instrument is in for a drastic overall in the Virginia class. For the first time an entire class of U.S. submarines is being fitted without all the prisms, mirrors, and lenses found in the old optical periscopes. In their place will be two non-hull-penetrating "photonics masts," which will consist of a number of high-resolution visual sensors that transmit visual images back to large display screens fitted within the boat. In addition to a color television pickup, there is also an advanced thermal and low-light imaging system. The photonics system will also

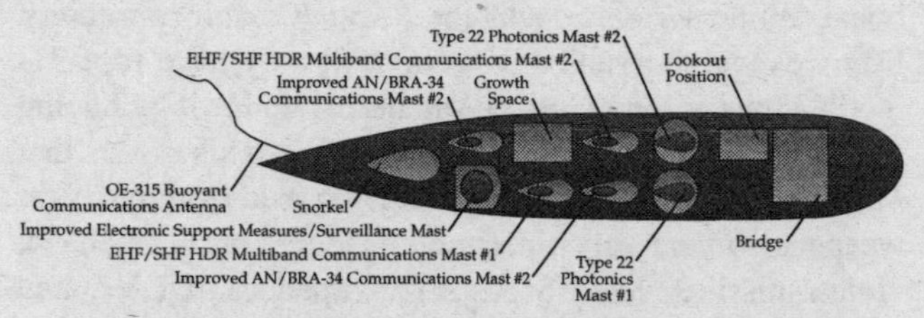

The arrangement of sensor and communications masts on the conning tower/sail of USS *Virginia* (SSN-774). *Rubicon, Inc., by Laura DeNinno*

contain a laser range finder, something that will come in handy while working in the close-in coastal regions. Best of all, as their name indicates, these masts do not penetrate the main pressure hull, making one less weak spot for water to leak through in the event of battle damage or a packing failure.

In addition, when the *Virginia* first enters service in 2004, she will likely be fitted with the new AN/BLQ-11A Long-term Mine Reconnaissance System (LMRS). The LMRS system is basically a team of UUVs (Unmanned Underwater Vehicles), with 21-inch/553mm diameter (so they fit in the torpedo tubes), that swim out and hunt freely for mines. These UUVs will not be connected to *Virginia* by a fiber-optic cable, as were earlier models. Instead, they will use a two-way acoustic data link. The UUVs can be launched and recovered (via a robotic arm in one of the torpedo tubes) from *Virginia*'s torpedo tubes and will detect and classify mines primarily of the bottom and moored variety. As the LMRS program continues to mature, additional improvements, such as underwater mapping and beach reconnaissance, will probably be added to this revolutionary UUV system.

Overall, the Virginias will have a weapons-storage capacity remarkably similar to that of the Seawolf-class boats. While *Virginia* lacks the *Seawolf*'s ability to carry fifty weapons internally (*Virginia* will only have four 21-inch/533mm torpedo tubes), she makes up for it by having twelve VLS tubes like those on the 688Is. This means that *Virginia* will have the ability to carry a total of thirty-eight weapons (twenty-six internally), including Tomahawk cruise missiles, Mk 48 ADCAP torpedoes, UUVs, and mines, to name just a few possibilities.

One other key element of the *Virginia*'s war-fighting suite is going to be her ability to operate in the special operations role. Similar to many of the newer submarines coming out of the yards in the next several years (most notably the *Jimmy Carter* [SSN-23]) the Virginias will be capable of carrying a Dry Dock Shelter and the new Advanced SEAL Delivery System (ASDS). The ASDS is a small mini-submarine 65 feet/21.7 meters long and 9 feet/2.75 meters wide. This tiny sub is fitted with a forward-looking and side-looking sonar to detect natural and man-made objects and conduct mine-detection and bottom-mapping operations. In addition, the ASDS has two masts—one a periscope and the other for communications and GPS navigation. To make life easier for the SEALs and other special operations units that might be carried, the torpedo room on board the Virginias will be easy to reconfigure. If need be, the center weapons-stowage structures can be removed in order to make room for special mission personnel, whether they be technicians, SEALs, or troops. These personnel will have access to an unusual nine-man lock-in/lockout chamber in addition to the use of the ASDS and/or dry-dock shelter.

With all these extra bodies on board, one begins to ask the next question: has submarine habitability improved in the past ten years? Unfortunately, advances in technology notwithstanding, the answer is likely to be no. Submarines have always been so crammed with equipment that it appears at times as if the designers forgot to include the sailors! Fortunately, numerous space-saving features have been included in the Virginia class. The current manning figures appear to be in the same neighborhood as the Los Angeles and Seawolf classes—around 120 enlisted and chiefs, along with 14 officers. With the significantly smaller size of the Virginias, this large complement could pose a problem. However, the manning numbers may end up dropping significantly as a result of the use of the Navy's new "Smart-Ship" technology.[7]

Already the Navy is planning on using this manpower-saving technology to reduce fifteen crew watchstanders from the control room of *Virginia*. This technology will allow the boat's advanced control system to be operated by a pilot, copilot, and relief pilot, thereby replacing the diving officer, chief of the watch, helmsman, planesman, and messenger of previous submarine classes. It is technologies such as these that will inevitably bring down the number of sailors to a more "comfortable" level. Alas, as any submariner knows, as soon as they get more room in a sub, designers will add more gadgets, not more racks!

Well, now that we've taken a look at the *Virginia*, it's pretty apparent that a boat like this can be used for virtu-

---

[7] First demonstrated aboard the Aegis cruiser *Yorktown* (CG-48), "Smart Ship" uses COTS-based systems to provide improved situational awareness to a reduced watch of crew personnel. It also helps with management of logistics and systems control, making the reduction of crew numbers a reality for the first time since the advent of steam.

ally any mission the Navy might require. Whether it's CVBG support in the Aegean, monitoring an embargo or communications in the Persian Gulf, delivering SOF units in Africa, or hunting enemy submarines and ships anywhere else, Virginia is one class of submarine that should be able to do it all! Now America just needs to build them, which may be the biggest challenge of all.

The British A-Class (Astute) Boats

While the United States Navy was going through its post–Cold War "shrinking pains," what was happening to the Royal Navy (RN) of our closest military ally, the United Kingdom? Well, you can rest assured that the RN submarine force was going through some equally painful experiences of its own! Like the United States, the British were in the middle of planning for the future of the Cold War when the Soviet Union threw in the towel at the end of 1991. You would have thought that when the Berlin Wall fell, it landed right on top of the Admiralty in London. As late as 1987, the British had been discussing a new class of attack boat to begin replacing the Swiftsure-class (S-126) SSNs. In 1987, Vickers Shipbuilding and Engineering (VSEL), LTD, won a contract to begin designing a new "W" class, also known as SSN-20, and was to start construction in the mid-1990s. This new submarine was similar in design and concept to the American *Seawolf*, large and optimized for blue-water/open-ocean operations.

The 1990s were to have been one of the busiest and most aggressive decades for British submarine development. However, while design work on the Swiftsure's replacement continued at VSEL, the RN was also in the midst of replacing its Resolution-class (S-22) ballistic missile submarines with a vaunted new group of SSBNs known as the Vanguard (S-28) class. While these truly awesome boats were in production, readying for their commissioning, the Upholder-class (S-40) SSKs were also scheduled to come into service during what would turn out to be one of the worst possible times imaginable. This class, consisting of *Upholder* (S-40), *Unseen* (S-41), *Ursula* (S-42), and *Unicorn* (S-43), were commissioned into service between 1990 and 1993, but by 1992 the decision was made to decommission all four of the brand-new SSKs as a cost-cutting measure.

The Upholder boats were eventually leased to Canada, but 1992 was a particularly dark time for the Royal submarine fleet. Also in that year, the entire SSN-20/W-class project to replace the Swiftsure was canceled with VSEL. Within months, however, a seed of hope was planted when the RN and Ministry of Defense (MoD) realized (as did their counterparts in the United States) that cost and *not* advanced blue-water operational capabilities was going to be the driving factor for getting a new class of SSNs ordered. Once this fact became apparent, priority switched from a new class of submarines to an improved version of the already proven Trafalgar (S-107) class SSNs.

British submarine designers were asked to submit a plan for a less expensive generation of submarines than had originally been envisioned. This new submarine was to vary only minimally from the previous Trafalgar-class SSNs. The boats, which soon took on the name Batch 2 Trafalgar class or B2TC, began to look like a reality when a request for bids was issued to both VSEL and competitor

GEC-Marconi in July of 1994. GEC ended up winning this contract, which called for construction of three boats with the option for more as deemed necessary and affordable by the MoD.

Initial progress was not exactly rapid, as complex defense consolidation issues stalled efforts to finalize the contract. The first actual orders for the B2TC boats were placed in March of 1997. The £2 billion contract called for three new boats, HMS *Astute, Ambush,* and *Artful*. Finally, the B2TCs had a real name—the "A" or Astute class of nuclear-powered attack submarines. The current outlook for the Astute class is excellent, based on both the needs of the Royal Navy and the design expertise of the manufacturer. The keel of the *Astute* herself was officially laid down in January 2001. Thankfully, MoD support has continued unabated for Astute. In fact, it was announced in July 1998 that the Royal Navy would get two more Astutes, for a class total of five boats. Though these two most recent boats have yet to be named, you can bet that their futures will be full of adventure.

While the busy process of finding a suitable replacement for the Swiftsures was under way, an equally complex dance of consolidation was going on within the British defense and shipbuilding industries. In 1994, GEC-Marconi made an offer to buy VSEL, which had been working on the original W-class/Swiftsure replacement designs. The following year, British Aerospace also made an offer for VSEL, but in the end VSEL accepted the GEC-Marconi offer—though only after they had successfully raised the asking price several hundred million pounds! As you might imagine, this caused considerable consternation, as VSEL had been in direct competition against GEC-Marconi to build the new Astutes. These issues were finally resolved, however, and all was quiet on the submarine industrial front for several more years. This lasted until 1999, when it was announced that British Aerospace was merging with GEC-Marconi. This meant that British Aerospace now took over construction of the new Astutes. In doing so, and as a result of its merger with GEC-Marconi, British Aerospace officially changed its name to BAE Systems and continued with *Astute*'s construction. Because of this, HMS *Astute*'s prime contractor has been referred to as VSEL, GEC-Marconi, British Aerospace, and BAE Systems. The free market of the "New World Order"—what a crazy world we live in!

Now let's walk down the gangplank and take a virtual look inside this new warship. The A-class boats will be based on a design concept similar to that of the Virginia class. That is to say, the Astutes will need to maintain the capability to fight in nearly every underwater environment, from the open ocean of the Atlantic to the littoral zones off the Persian Gulf, all while keeping costs down and weapons loads and capabilities up. From protection of the British homeland to far-off TLAM attacks, the missions of *Astute* require that she contain only the best electronics and weapons available.

One of the biggest differences between Trafalgar and Astute is the size of the boats' hulls. Astute will displace around 7,000 tons submerged and 6,390 tons surfaced compared with 5,208 tons and 4,740 tons for the Trafalgar-class SSNs. Much of this tonnage, along with Astute's longer length (318 feet/97 meters compared with 279 feet/85 meters of the T-class boats—an increase of more than 36 feet/10 meters!), can be attributed to the newer boat's larger weapons load-out.

Where the Trafalgar-class boats carried a total of twenty-five weapons, the new warships will be able to hold thirty-six (at least that's the number officially being claimed). They will be fired by six torpedo tubes versus the five in the earlier class. Some consideration was given to providing the Astute-class SSNs with a VLS system, but in light of cost constraints, the new boats' warload was deemed sufficient.

Another key update included in Astute has been her reactor plant. Astute's power plant was originally developed for the Vanguard-class SSBN. Designated PWR 2, it will be built by Rolls-Royce and is rated at 15,000 shaft horsepower. However, the Astute version is significantly more advanced than the earlier variants, and some press reports speculate that the new power plant is capable of circling the globe literally scores of times without a single refueling. Reportedly there is a lifetime reactor core design (known as an "H" core) that will give it twenty-five- to thirty-year service life. According to the Royal Navy submarine fleet's current plan, by the time refueling is needed, the first Astute will hopefully be resting her keel on the front porch of a submarine retirement community!

Last but certainly not least, Astute's propulsion systems would not be complete without the two turbines with a single shaft connected to the now standard (and *very* quiet) pumpjet propulsor. Of course, quieting is the most important characteristic for submarine survival in a hostile environment these days, and Astute's improved reactor will be significantly quieter than Trafalgar's. As a whole, Astute will be capable of operating not just in littoral and blue-water regions, but also in such varied climates as those that are found under the arctic ice or in the warm waters of the tropics.

In the control room of submarines everywhere, processing and computer power has also been increasing at an astounding rate for decades now. Thus any new class of boats, even improved models such as Astute, will have its share of both updated legacy systems along with new gadgets and sensors. Radar and sonar systems for the Astutes will be essentially the same units as were found in the Trafalgars, albeit with some improvements. It now appears that the Astutes will be fitted with the same Type 2076 sonar system, which has already been successfully integrated into the most recent Trafalgar-class boats. It is also all but certain that the *Astute* will be fitted with towed array sonar, as have her predecessors, although the specific variant and configuration have yet to be announced. Finally, current plans call for the Astute class to contain her own non-hull-penetrating "optronics" masts with all-electronic displays, minus the lenses, mirrors, and prisms of traditional periscopes. These are similar to those that will be carried on *Virginia* and will likely provide some brand-new capabilities and challenges for the Perisher-trained skippers of the RN.

What is a submarine without its weapons? RN attack submarines have now been fitted with Block III Tomahawk cruise missiles, and as the world learned during NATO's Operation Allied Force in 1999, Royal Navy submarines can shoot a lot of them, too! HMS *Splendid* launched more than two dozen against heavily defended Serbian targets, a sizable portion of the limited supply purchased from the U.S. When *Astute* enters service in June 2005, she will be equipped with a sizable complement of TLAMs, and the Royal Navy is also looking at development of a tube-launched version of the TACTOM missile currently under development for the U.S. Navy.

Land attack missiles aside, Astute also has the requirement to engage surface ships and other submarines. For this mission, Astute's six torpedo tubes will be compatible with both the RN version of the sea-skimming UGM-84 Block 1C Harpoon missile (for attacks against surface vessels) and the fast, albeit expensive, wire-guided Spearfish torpedo (for use against both submarines and surface ships). Mines can also be carried and may have a particular use in the littoral regions.

Habitability and sustainability are always key issues on a nuclear submarine. When judging a submarine with the capability of spending extended periods of time away from port and beneath the surface, habitability must always be taken into consideration. Current plans call for a crew of twelve officers and ninety-seven enlisted sailors aboard *Astute.* While increased size has led to the improvement of some accommodations compared with the Trafalgars, some Astute sailors will still probably have to hot bunk with their shipmates. This practice will be required for only the most junior of her sailors, since current estimates call for the sharing of around eighteen of the crew bunks.

All of the improvements and modernizations built into the Astutes will become a reality when the first of the class is commissioned into service in 2006, followed by *Ambush* in 2007 and *Artful* in 2008. As mentioned earlier, it now appears that at least two additional A-class boats will be built, and possibly even more, though these have yet to be named or officially ordered.

What comes next for the Royal Navy's submarine force? Work has already begun on a submarine to replace the initial seven Trafalgar boats. The boats are so far unnamed, but what is known is that they will be nuclear powered and have a submerged displacement of between 5,000 and 8,000 tons. They will also be fitted with a "life of the boat" nuclear reactor, similar to those found in the Astutes and Virginias. In addition, VLS tubes may be fitted to this next class to allow for an increased load-out of Tomahawk Block III or TACTOM land attack missiles. As with the Virginias, any new attack submarine the Royal Navy builds will likely have provisions for UUVs. There is even talk that the Astutes will be the last manned submarine to enter service with the Royal Navy, something that requires more than a minor leap of faith and vision in the future of technology. In the end, though, the real strength of the RN SSN force will be what it always has been—superbly trained crews and Perisher-qualified skippers, able to outguess and outgun their enemies.

Conclusions: Toward the Unknown

Yes, a decade really does make a difference. Ten years ago we asked what the next decade of submarine development would be like. Well, after reading this last chapter, you can see that question has clearly been answered. Fortunately, the world has also learned some lessons about the nature of naval warfare along the way. Most importantly, we've learned that if there's one thing we can't predict, it's the future! No one could have accurately predicted the end of the Cold War, and likewise no one will ever be able to accurately predict exactly what the next generation of submarine operations and missions will be like.

We do know that the past ten years of submarine operations have consisted of the unexpected. From Operation Desert Storm to our Navy's involvement in the recent

NATO actions in the Balkans, the one theme that has remained the same throughout the 1990s has been *doing more with less.* The submarine communities of the U.S. and Royal navies, along with those of our other allies, have had their budgets cut, their submarines retired from service, and their crews downsized, all while being asked to conduct more activities and operations than at any time in recent memory. All this has been happening while highly capable submarines such as the Project 877/Kilo-class boats have been entering the navies of "rogue" nations at a far greater pace than new classes such as Seawolf, Virginia, and Astute will be entering ours.

What's the solution to this problem? In today's high-threat environment, where we don't have the benefit of looking in a specific direction to watch for flying bullets, we must be prepared for every contingency. And we are. A quick look at America's modern submarine force shows that. They're out there on patrol for the nation, in the backyards of our enemies and competitors—literally on their doorstep day in and day out—*already prepared for any contingency*. Whether it's a TLAM strike against a future enemy nation or a special operations rescue mission launched from the depths of a submarine dry-dock shelter, the sailors of today's submarine fleet are doing their best to stay prepared for *any* emergency. And for that we owe them our respect.

Getting back to our first question, what exactly does the future hold for the submarine force? The truth is, we don't know. Maybe the Virginias and the Astutes will be the last manned submarines the U.S. and Britain will ever build, eventually to be replaced by advanced unmanned submersibles. Or perhaps our fleets will become more and more sub-surface-centric as the surface ship becomes increasingly vulnerable to antiship missiles and other weapons. We just don't know yet what the future holds. However, whatever it may be, our nation's submariners (hopefully along with those of our allies) will be on patrol, ready to defend our national interest in times of trouble and threat. Can we ask for anything more?

Other People's Submarines

It is something of an oddity that in a world where the numbers and sizes of military forces are decreasing, submarines continue to be built. In fact, while it is not quite a growth industry, production of diesel-electric submarines is continuing in a number of countries and yards worldwide. In addition, those countries that have the capability to build nuclear submarines are fighting desperately to maintain the industrial base to do so. While the nations of the world have been downsizing their own submarine forces, they have also been trying to market the products of their building yards to developing countries that desire an entry into the world of submarine capabilities.

It also is rather ironic that while the number of submarines in the world has decreased radically, the overall quality and age of the remaining boats has improved—rather a strange situation for those who suggest that peace has broken out around the world. Thus, anyone choosing to hunt other submarines is facing the reality that the task is probably getting tougher. In addition, nations that are generally considered as outlaws (like Iran and Algeria) are obtaining a number of new production diesel-electric boats. This proliferation means that the United States and our allies may have to hunt enemy subs in places we have never gone before. The recent deployment of the USS *Topeka* (SSN-754) to the Persian Gulf, at the same time the first of the Iranian Kilo-class boats was being delivered, is probably not a coincidence. Even more interesting, though, would have been to see if another U.S. boat, perhaps even another 688I, was invisibly trailing the Kilo on its delivery trip.

The section that follows is a compendium of the more modern submarines, both nuclear and conventionally powered, that might face the U.S. sub force. Some of them, like those of the United Kingdom and France, are operated by nations that are considered allies. Others, like those of the Russian Navy and the clients of the Germans and the French, might still pose a threat to the forces allied with the United States. This should not be considered a list of every single boat in the world, however. For that, I defer to A. D. Baker's incomparable biannual work, *Combat Fleets of the World* (Naval Institute Press, Annapolis, Md.).

For the benefit of the reader, the following explanations of the terms used is provided:

CLASS NAME: Name of the first boat of the class or building program

PRODUCER (COUNTRY/MANUFACTURER): Country of origin and production site

DISPLACEMENT (SURFACED/SUBMERGED): Surfaced and submerged in long tons (2,240 lbs per)

DIMENSIONS (FT/M): Length: Bow to stern; **Beam:** Side to side; **Draft:** To keel

Armament: Number of tubes/launchers and weapons
Machinery: Power plant(s), number of screws, propeller blades, and shaft horsepower (SHP)
Speed (knots): Maximum
Number in class: In service + building + planned
Users: All countries currently using
Comments: Some thoughts and features of the class

Russia/Commonwealth of Independent States

While the "evil empire" of the Soviet Union may be dead, the navy built by the USSR is alive and still useful. Despite having suffered the breakup of the nation it was designed to serve, and having scrapped over half of all its warships, the Russian Navy is still one of the most powerful fighting forces afloat. They still deploy something like 240 submarines of various types, as well as a large array of surface ships. And while the Russian Navy and its CIS brother services are suffering from a shortage of almost everything, the missile boats are still making their deployments, with the attack boats still supporting them in the bastions.

The big challenge for the Russian submarine force, like everyone else, is surviving the present to move on to the future. Their first problem is, of course, how to maintain their existing force of attack, guided missile, and ballistic missile submarines. This problem has been made extremely difficult by the financial troubles of the Russian Republic, though they have managed to hold things together until now. Another problem is the fleet of aged submarines (many of them nuclear powered) that they have. The recent news photograph of an abandoned Russian submarine poking its bow through the winter pack ice in

Vladivostok is a chilling statement on the inability of the Russians to deal with this problem. Clearly, just how to dispose of over 150 obsolete nuclear submarines is a problem that will require the help of the United States and her allies.

As for the future, only events will tell us that. The one thing that does seem certain is that the Russians will continue development of submarines and their related technology. While many of the design bureaus for aircraft and tanks are in desperate trouble, there are continuing reports that they are still funneling their limited military R&D funds into designs for newer and quieter submarines. The most likely projects for such research will probably be a replacement for the Delta IV SSBNs, a new SSN derived from the highly successful Akula-class boats, and possibly a new diesel-electric design to replace the Kilo and support export sales. The replacement SSBN makes sense in light of the new START arms agreements, which have the Russians placing over half their deliverable nuclear warheads on submarine-launched missiles. And just as obviously, the Akula and Kilo replacements will be needed to protect those SSBNs and maintain the credibility of the CIS nuclear deterrent.

Overall, this is a major reduction from several years ago, when the R&D effort was probably two to three times this size. Rumor had the Russians working on replacements for the Oscar-class SSGNs, the Typhoon SSBNs, the SSNs of the Sierra class, and even a Rubis-sized SSN for export to the Indian Navy. All of this though is based on what we see happening today. And as any honest watcher of Russian military trends will tell you, the crystal ball is cloudy and the tea leaves unreliable where they are concerned. In the end, it will probably come down to whether Boris Yeltsin can hold things together long enough for an actual trend to develop. So here it is as of today.

Victor III. *Jack Ryan Enterprises, Ltd.*

CLASS NAME: Victor III (Project 671 RTM)

PRODUCER (COUNTRY/MANUFACTURER): Russia/Russian Admiralty; Komsomolsk

DISPLACEMENT (SURFACED/SUBMERGED): 4,900/6,000

DIMENSIONS (FT/M): Length: 341.1/104 **Beam:** 32.8/10 **Draft:** 23/7

ARMAMENT: Four 650mm and two 533mm torpedo tubes with 24 weapons

MACHINERY: Two PWRs with steam turbines driving one tandem 8-bladed screw; 30,000 SHP

SPEED (KNOTS): 30 (submerged)

NUMBER IN CLASS: 26

USERS: Russia

COMMENTS: While it soon will be the oldest class of SSN in the Russian inventory, the Victor III is still a dangerous and capable opponent. Well armed and relatively quiet (roughly similar to the Sturgeon class), this boat was the first Soviet SSN capable of matching Western boats. The stern pod, first found on Victor IIIs, is now a feature of every modern Russian SSN; it contains a passive towed array sonar system.

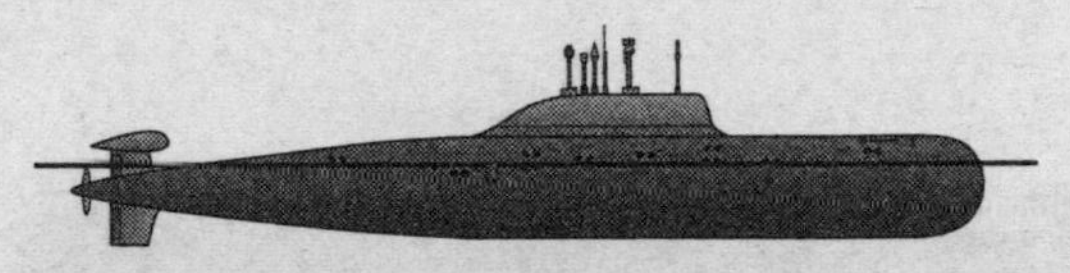

Akula. *Jack Ryan Enterprises, Ltd.*

CLASS NAME: Akula (Russian: *Bars* class) (Project 971)

PRODUCER (COUNTRY/MANUFACTURER): Russia/Severodvinsk, Komsomolsk

DISPLACEMENT (SURFACED/SUBMERGED): 7,500/10,000

DIMENSIONS (FT/M): Length: 370.6/113 **Beam:** 42.6/13 **Draft:** 32.8/10

ARMAMENT: Four 650mm and four 533mm torpedo tubes with 30+ weapons

MACHINERY: Two PWRs with steam turbines driving one 7-bladed screw; 45,000 SHP

SPEED (KNOTS): 35 (submerged)

NUMBER IN CLASS: 7+?

USERS: Russia

COMMENTS: When western submariners have nightmares, they usually revolve around this class of SSN. Akula is the quietest SSN yet produced by Russia and represents a boat in the class of a Flight I Los Angeles. Probably utilizes a raft sound isolation system to keep noise down. Reportedly the last remaining Russian SSN class still in production. President Yeltsin has announced that the Komsomolsk shipyard, located in the far east, is going to be converted to civilian production by 1995 or '96. After that, only the Severodvinsk shipyard, located in the Kola Peninsula, will produce submarines for the Russian Navy.

Sierra. *Jack Ryan Enterprises, Ltd.*

Class name: Sierra I/II (Russian: *Barrakuda* class) (Project 945A and 945B)

Producer (country/manufacturer): Russia/Krasnaya Sormova

Displacement (surfaced/submerged): Sierra I—6,050/7,600; Sierra II—6,350/7,900

Dimensions (ft/m): Length: 351/107 or 367.4/112 **Beam:** 41/12.5 **Draft:** 24.3/7.4

Armament: Four 650mm and two 533mm torpedo tubes with an estimated 30 weapons

Machinery: Two PWRs with steam turbines driving one 7-bladed screw; 45,000 SHP

Speed (knots): 35 (submerged)

Number in class: 2/1 + 1

Users: Russia

Comments: The evolutionary descendant of the Alfa, the Sierra is a titanium-hulled follow-on to the previous classes of Soviet SSNs. Very quiet and well armed, it has been overshadowed by the highly successful steel-hulled Akula-class boats. Reportedly, once the last Sierra II is completed, Krasnaya Sormova will convert to civilian ship production.

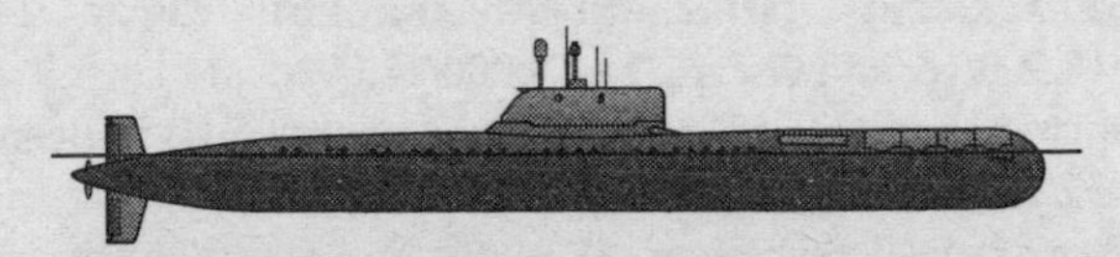

Charlie II. *Jack Ryan Enterprises, Ltd.*

CLASS NAME: Charlie II (Project 670M)

PRODUCER (COUNTRY/MANUFACTURER): Russia/Krasnaya Sormova

DISPLACEMENT (SURFACED/SUBMERGED): 4,300/5,500

DIMENSIONS (FT/M): Length: 337.8/103 **Beam:** 32.8/10 **Draft:** 26.2/8

ARMAMENT: Eight SS-N-9s in external tubes; six 533mm torpedo tubes with 12 weapons

MACHINERY: One PWR with steam turbines driving one 5-bladed screw; 15,000 SHP

SPEED (KNOTS): 24 (submerged)

NUMBER IN CLASS: 6

USERS: Russia

COMMENTS: These boats may possibly be the oldest guided missile submarines that will be retained by Russia. Relatively noisy, but they can still pack a powerful punch with their battery of SS-N-9 Siren antiship missiles.

Oscar. *Jack Ryan Enterprises, Ltd.*

CLASS NAME: Oscar I/II (Russian: *Granite/Antey* classes) (Project 949 & 949A)

PRODUCER (COUNTRY/MANUFACTURER): Russia/Severodvinsk

Displacement (surfaced/submerged): Oscar I—13,900/16,700; Oscar; II—15,000/18,000

Dimensions (ft/m): Length: 478.9/146 or 505.1/154 **Beam:** 59/18 **Draft:** 32.8/10

Armament: Twenty-four SS-N-19s in external tubes; six 650mm and 533mm torpedo tubes with 24 weapons

Machinery: Two PWRs with steam turbines driving two 7-bladed screws; 90,000 SHP

Speed (knots): 33 (submerged)

Number in class: 2/7+

Users: Russia

Comments: *Oscar* carries the nickname of "Mango" for her size and firepower. She also is as quiet as Sierra and carries the same sonar system, including a towed array from the tube on top of the rudder, as the Sierra-class boats. With 24 SS-N-19 Shipwreck heavy antiship missiles and a full array of torpedoes, this is the largest and most heavily armed attack submarine in the world. Probably capable of taking one or more torpedo hits and still surviving.

Class name: Fourth generation SSN (Akula follow-on) (Project: ?)

Producer (country/manufacturer): Russia/Severodvinsk

Displacement (surfaced/submerged): ≈ 10,000 (submerged)

Dimensions (ft/m): (unknown)

Armament: Six to eight 650mm and 533mm torpedo tubes with 30+ weapons

Machinery: PWRs with steam turbines driving one 7-bladed screw; ? SHP

Speed (knots): ≈ 30–35 (submerged)

Number in class: ?
Users: Russia
Comments: If the Russians choose to continue SSN production, they will probably base the fourth-generation design on their formidable Akula class. In terms of capability, this boat will probably be the equal of a 688I in terms of quieting, and will have improvements in sonar, computers, and weapons. Should a decision be made to produce it, the first boat will probably be commissioned somewhere in the 2003–2005 time frame.

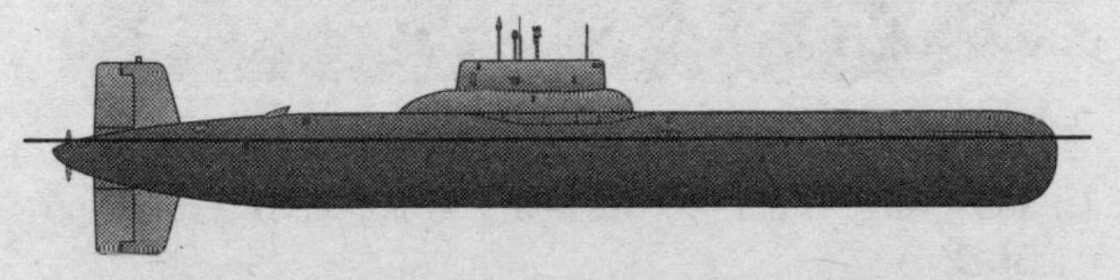

Typhoon. *Jack Ryan Enterprises, Ltd.*

Class name: Typhoon (Russian: *Akula* class) (Project 941)
Producer (country/manufacturer): Russia/Severodvinsk
Displacement (surfaced/submerged): 18,500/25,000
Dimensions (ft/m): Length: 560.9/171 **Beam:** 78.7/24 **Draft:** 41/12.5
Armament: Twenty SS-N-20 SLBMs; six 650mm and 533mm torpedo tubes with an estimated 24 weapons
Machinery: Two PWRs with steam turbines driving two shrouded 7-bladed screws; 90,000 SHP
Speed (knots): 25 (submerged)
Number in class: 6
Users: Russian-operated but under CIS control
Comments: The world's biggest submarine, pure and simple. The Typhoon seems to have been designed as a direct counter to the Ohio-class SSBNs. She carries 20 equally huge SS-N-20 (RSM-52) Sturgeon submarine-launched

ballistic missiles. Essentially two Delta IV pressure hulls slapped together, with additional spaces for torpedo tubes and storage and ship control, this monster of the deep is equipped for long-term operations, particularly in the Arctic areas. Because of its double hull and massive bulk, to sink it with just a single heavy torpedo hit would be nearly impossible. The Russians call this beast *Akula*.

Delta IV. *Jack Ryan Enterprises, Ltd.*

CLASS NAME: Delta IV (Russian: *Del'fin* class) (Project 667 BRDM)

PRODUCER (COUNTRY/MANUFACTURER): Russia/Severodvinsk

DISPLACEMENT (SURFACED/SUBMERGED): 10,800/13,500

DIMENSIONS (FT/M): Length: 537.9/164 **Beam:** 39.4/12 **Draft:** 28.5/8.7

ARMAMENT: Sixteen SS-N-23 SLBMs; six 650mm and 533mm torpedo tubes with 18 weapons

MACHINERY: Two PWRs with steam turbines driving two 7-bladed screws; 50,000 SHP

SPEED (KNOTS): 24 (submerged)

NUMBER IN CLASS: 7

USERS: Russian-operated but under CIS control

COMMENTS: A direct descendant of the highly successful Delta III SSBNs, the Delta IV was originally seen as a "just in case" program should the Typhoons not perform as desired. Unveiled in the light of the START II world, this formidable design has been seen for what it is, a *very* capable and quiet SSBN able to sustain long operations, even in the Arctic regions. Delta IVs carry 16 liquid-fueled SS-N-23 (RSM-54) Skiff submarine-launched ballistic missiles.

Delta III. *Jack Ryan Enterprises, Ltd.*

CLASS NAME: Delta III (Russian: *Kal'mar* class) (Project 667 BDR)

PRODUCER (COUNTRY/MANUFACTURER): Russia/Severodvinsk

DISPLACEMENT (SURFACED/SUBMERGED): 10,600/13,250

DIMENSIONS (FT/M): Length: 510/155.5 **Beam:** 39.4/12.0 **Draft:** 28.2/8.6

ARMAMENT: Sixteen SS-N-18 SLBMs; six 533mm torpedo tubes with 18 weapons

MACHINERY: Two PWRs with steam turbines driving two 5-bladed screws; 50,000 SHP

SPEED (KNOTS): 24 (submerged)

NUMBER IN CLASS: 14

USERS: Russian-operated but under CIS control

COMMENTS: First appearing in the mid-1970s, the Delta IIIs were the first Soviet SSBNs to truly rival the American SSBNs in weapons. Specifically, her long-range SS-N-18 (RSM-50) Stingray missiles with multiple reentry vehicles gave her the ability to hit numerous targets in North America from pier side at either Petropavlovsk or Murmansk bases. Delta IIIs will probably be the oldest SSBNs retained by the Russian Navy under START II, and some will probably serve until the beginning of the twenty-first century.

CLASS NAME: Fourth-generation SSBN (Delta IV follow-on) (Project: ?)

PRODUCER (COUNTRY/MANUFACTURER): Russia/Severodvinsk

DISPLACEMENT (SURFACED/SUBMERGED): ≈ 13,000–15,000 (submerged)

DIMENSIONS (FT/M): (unknown)

ARMAMENT: est. sixteen SS-N-? SLBMs; six 650mm and 533mm torpedo tubes with about 20 weapons

MACHINERY: PWRs with steam turbines driving two 7-bladed screws; ? SHP

SPEED (KNOTS): ≈ 25–30 (submerged)

NUMBER IN CLASS: ?

USERS: Russia

COMMENTS: Should the Russians choose to build a fourth-generation SSBN, it will probably be based upon the highly successful Delta IV boat. Improvements will probably be limited to further quieting and some refinement to weapons (improved targeting and accuracy), as well as sensors.

Kilo. *Jack Ryan Enterprises, Ltd.*

CLASS NAME: Kilo (Russian: *Varshavyanka* class) (Project 877)

PRODUCER (COUNTRY/MANUFACTURER): Russia/Komsomolsk, Krasnaya Sormova, United Admiralty

DISPLACEMENT (SURFACED/SUBMERGED): 2,325/3,076

DIMENSIONS (FT/M): **Length:** 243.7/74.3 **Beam:** 32.8/10 **Draft:** 21.6/6.6

ARMAMENT: Six 533mm torpedo tubes with 18 weapons
MACHINERY: Diesel-electric drive with one 6-bladed screw; 5,900 SHP
SPEED (KNOTS): 17 (submerged)
NUMBER IN CLASS: 20+
USERS: Russia, Poland, Algeria, Romania, India, Iran
COMMENTS: Currently the only diesel-electric submarine known to be in production in Russia. The Kilo is a medium-sized, inexpensive SSK with excellent quieting and weapons, although the lack of a towed array limits it in the area of sensors. Something of a best-seller, the Kilo has become an important source of hard currency for the struggling Russian sub builders, though newer, more advanced Western designs may intrude on sales.

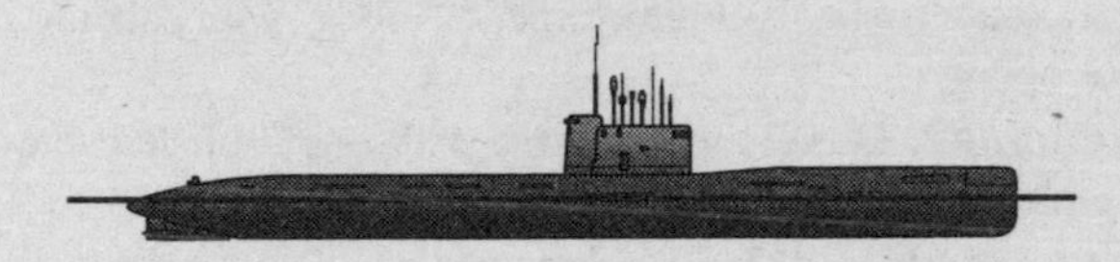

Tango. *Jack Ryan Enterprises, Ltd.*

CLASS NAME: Tango (Project 641B)
PRODUCER (COUNTRY/MANUFACTURER): Russia/Krasnaya Sormova
DISPLACEMENT (SURFACED/SUBMERGED): 3,100/3,900
DIMENSIONS (FT/M): Length: 300.1/91.5 **Beam:** 29.5/9 **Draft:** 23/7
ARMAMENT: Ten 533mm torpedo tubes with 24 weapons
MACHINERY: Diesel-electric drive with three 5-bladed screws; 6,000 SHP
SPEED (KNOTS): 20 (submerged)
NUMBER IN CLASS: 18
USERS: Russia

Comments: One of the last *big* classes of diesel boat to be constructed by the Soviet Union, the Tangos were originally designed to be open ocean SSKs with the primary missions of attacking aircraft carriers and interdicting merchant ships. Extremely quiet and capable, they have excellent range and weapons. A number of the Tangos will continue to serve until the turn of the century.

Class name: Fourth-generation SS (Kilo follow-on) (Project: ?)

Producer (country/manufacturer): Russia/Severodvinsk

Displacement (surfaced/submerged): ≈ 2,500–3,000 (submerged)

Dimensions (ft/m): (unknown)

Armament: six 533mm torpedo tubes with about 20 weapons

Machinery: Diesel-electric drive with one 7-bladed screw; probable AIP system; ? SHP

Speed (knots): ≈ 25–30 (submerged)

Number in class: ?

Users: Russia and ?

Comments: If Russia chooses to keep building conventionally powered submarines, they will probably base the next design on a prototype boat known as Beluga that has been undergoing testing in the Black Sea. The new design SS may utilize a novel Air Independent Propulsion (AIP) system to extend slow-speed submerged endurance, which reduces the time the submarine would have to spend snorkeling. In addition, with a hull form based on the Alfa-class SSNs the fourth-generation SS may be capable of high SSN-like speeds for short periods.

PEOPLE'S REPUBLIC OF CHINA

While the Russians ran headlong into the business of building nuclear submarines, the People's Republic of China (PRC) took a slow, steady pace. Their first SSN, the Han class, is a simple boat with very little of the high technology that would be considered standard on an American or British boat. From the Hans has come the Xia class, the PRC's first SSBN. It appears that both the Han and the Xia have finished their production runs. With only six first-generation units, the Chinese appear to have mixed feelings about the success of the Han and Xia. Nevertheless, it is likely that within the foreseeable future, the Chinese will begin production of the Han and Xia follow-ons.

CLASS NAME: Han

PRODUCER (COUNTRY/MANUFACTURER): PRC/Huludao

DISPLACEMENT (SURFACED/SUBMERGED): 4,500 (submerged)

DIMENSIONS (FT/M): Length: 295.2/90 **Beam:** 32.8/10 **Draft:** ?

ARMAMENT: Six 533mm torpedo tubes

MACHINERY: One PWR with turboelectric drive; one-bladed screw; 15,000 SHP

SPEED (KNOTS): 30 (submerged)

NUMBER IN CLASS: 5

USERS: PRC

COMMENTS: This is the first class of SSN produced by the PRC, and it shows. Rather noisy and limited in its weapons load and sensors, it is still relatively fast and better than nothing. Roughly equal to a Skipjack or Victor I in performance.

Class name: Xia
Producer (country/manufacturer): PRC/Huludao
Displacement (surfaced/submerged): 7,000 (submerged)
Dimensions (ft/m): Length: 393.6/120 **Beam:** 32.8/10 **Draft:** ?
Armament: Twelve CSS-N-3 SLBMs; six 533mm torpedo tubes
Machinery: One PWR with turboelectric drive; one-bladed screw; 15,000 SHP
Speed (knots): 20 (submerged)
Number in class: 1
Users: PRC
Comments: The first class of SSBNs built by the PRC, the Xias are roughly similar to the Soviet Yankee II-class boat in performance and weapons loadout. A derivative of the Han class (the hull and reactor are virtually identical), the Xias give the PRC leadership a minor and somewhat credible FBM capability in their part of the world.

France

The French are somewhat unusual because they chose to develop nuclear submarines to carry ballistic missiles (SSBNs) before they developed nuclear attack submarines (SSNs). This was due to the desire of General Charles de Gaulle in the 1960s to have a nuclear deterrent independent of NATO. Consequently, they developed a force of four SSBNs, the Le Redoubtable class, first, and only recently developed a force of SSNs. Currently they are finishing the construction of the Améthyst-class SSNs, as well as working on a new class of

SSBNs, the four units of the Le Triomphant class. In addition, the French maintain a small force of diesel-electric submarines, though the number of these will surely decrease. As for the future, the French plans are not clear beyond the attempts of commercial yards to market conventional versions of the Améthyst-class boats.

Améthyst (French). *Jack Ryan Enterprises, Ltd.*

Class name: Améthyst

Producer (country/manufacturer): France/DCAN, Cherbourg

Displacement (surfaced/submerged): 2,400/2,660

Dimensions (ft/m): Length: 241.4/73.6 **Beam:** 24.9/7.6 **Draft:** 21/6.4

Armament: Four 533mm torpedo tubes with 14 weapons

Machinery: One PWR with turboelectric drive; one 7-bladed screw; 9,500 SHP

Speed (knots): 28 (submerged)

Number in class: 1 + 2

Users: France

Comments: Basically an improved Rubis with a rounded bow, these boats are superior in both radiated noise and sensors. Currently there are no plans by the French Navy for a larger SSN force, and these will be the last ones built in the foreseeable future.

Rubis (French). *Jack Ryan Enterprises, Ltd.*

CLASS NAME: Rubis
PRODUCER (COUNTRY/MANUFACTURER): France/DCAN, Cherbourg
DISPLACEMENT (SURFACED/SUBMERGED): 2,385/2,670
DIMENSIONS (FT/M): Length: 236.5/72.1 **Beam:** 24.9/7.6 **Draft:** 21/6.4
ARMAMENT: Four 533mm torpedo tubes with 14 weapons
MACHINERY: One PWR with turboelectric drive; one 7-bladed screw; 9,500 SHP
SPEED (KNOTS): 25 (submerged)
NUMBER IN CLASS: 4
USERS: France
COMMENTS: The first of the French SSNs, the Rubis class appeared only in the last decade or so. These compact little boats are the smallest SSNs ever built, and this is reflected in the small crew size (8 officers and 57 men) and the weapons loadout (14 weapons). The early units of the class were reported to be relatively noisy, requiring a major refit. All units are currently being brought up to the standard of the Améthyst-class boats.

CLASS NAME: Le Triomphant
PRODUCER (COUNTRY/MANUFACTURER): France/DCAN, Cherbourg
DISPLACEMENT (SURFACED/SUBMERGED): 12,640/14,120

Dimensions (ft/m): Length: 452.6/138 **Beam:** 41/12.5 **Draft:** ?
Armament: Sixteen M45 SLBMs; four 533mm torpedo tubes with ? weapons
Machinery: One PWR with steam turbines driving one pumpjet propulsor; 41,500 SHP
Speed (knots): 25+ (submerged)
Number in class: 1 + 3
Users: France
Comments: New generation of French SSBN. Considerable attention paid to quieting measures including main propulsion machinery raft and a pumpjet propulsor. More streamlined hull form over L'Inflexible and earlier SSBNs. Will be equipped with the latest in submarine sonar systems including a large flank array, and new combat system.

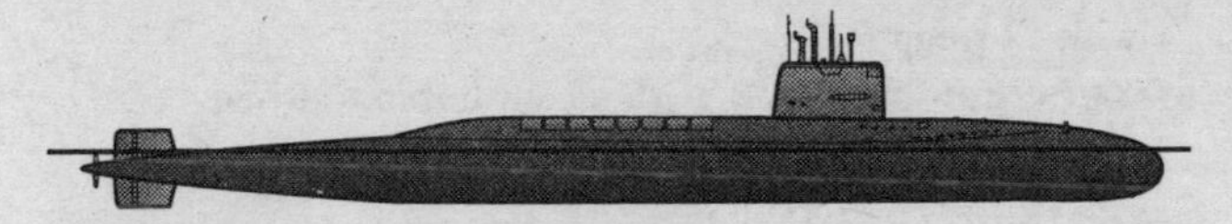

L'Inflexible (French). *Jack Ryan Enterprises, Ltd.*

Class name: L'Inflexible
Producer (country/manufacturer): France/DCAN, Cherbourg
Displacement (surfaced/submerged): 8,080/8,920
Dimensions (ft/m): Length: 422.1/128.7 **Beam:** 34.8/10.6 **Draft:** 32.8/10
Armament: Sixteen M4 SLBMs; four 533mm torpedo tubes with 12 weapons
Machinery: One PWR with steam turbines driving one 7-bladed screw; 16,000 SHP
Speed (knots): 20 (submerged)
Number in class: 1

Users: France

Comments: Essentially a Le Redoubtable class with some improvements in quieting, hull steel, and sensors.

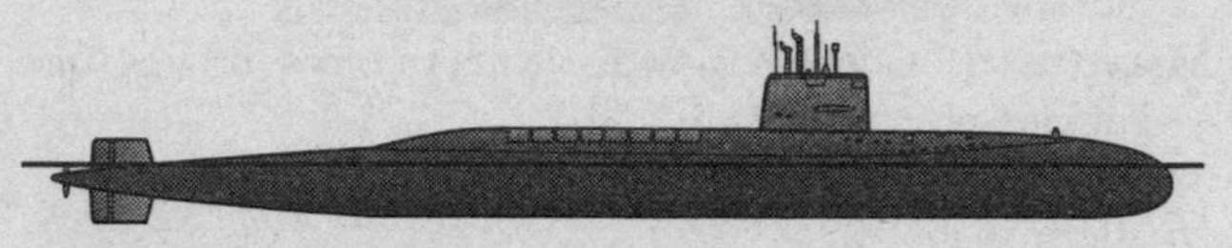

Le Redoubtable (French). *Jack Ryan Enterprises, Ltd.*

Class name: Le Redoubtable

Producer (country/manufacturer): France/DCAN, Cherbourg

Displacement (surfaced/submerged): 8,000/9,000

Dimensions (ft/m): Length: 419.8/128 **Beam:** 34.8/10.6 **Draft:** 32.8/10

Armament: Sixteen M4 SLBMs; four 533mm torpedo tubes with 12 weapons

Machinery: One PWR with steam turbines driving one 7-bladed screw; 16,000 SHP

Speed (knots): 20 (submerged)

Number in class: 4

Users: France

Comments: The first class of SSBNs constructed by the French Navy and, in fact, the first nuclear ship built indigenously in Europe. Le Redoubtable was decommissioned in December 1991; all others in the class undergoing modernization to L'Inflexible standard.

Agosta (French). *Jack Ryan Enterprises, Ltd.*

CLASS NAME: Agosta
PRODUCER (COUNTRY/MANUFACTURER): France/DCAN, Cherbourg
DISPLACEMENT (SURFACED/SUBMERGED): 1,490/1,740
DIMENSIONS (FT/M): Length: 221.7/67.6 **Beam:** 22.3/6.8 **Draft:** 17.7/5.4
ARMAMENT: Four 550mm torpedo tubes with 20 weapons
MACHINERY: Diesel-electric drive with one 7-bladed screw; 4,600 SHP
SPEED (KNOTS): 20 (submerged)
NUMBER IN CLASS: 4
USERS: France, Pakistan, Spain
COMMENTS: The last general purpose diesel-electric subs built by the French. An excellent design, they are being upgraded to the standards of the Améthyst class.

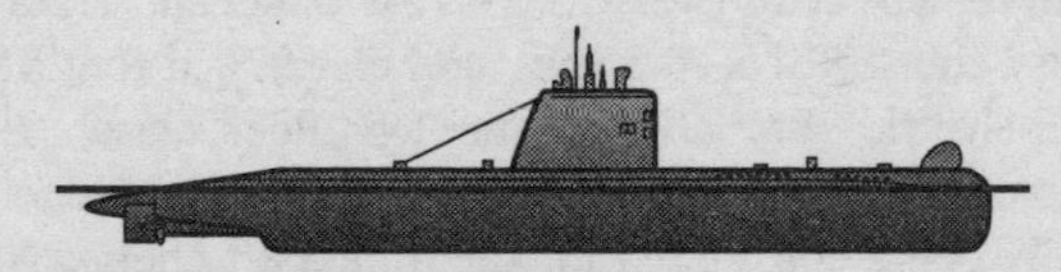

Dauphné (French). *Jack Ryan Enterprises, Ltd.*

CLASS NAME: Dauphné
PRODUCER (COUNTRY/MANUFACTURER): France/DCAN, Cherbourg
DISPLACEMENT: (SURFACED/SUBMERGED): 869/1,043
DIMENSIONS (FT/M): Length: 188.9/57.6 **Beam:** 22.2/6.8 **Draft:** 17.4/5.3
ARMAMENT: Twelve 550mm torpedo tubes with 12 weapons

Machinery: Diesel-electric drive with two 3-bladed screws; 2,000 SHP

Speed (knots): 16 (submerged)

Number in class: 19

Users: France, Pakistan, Portugal, Spain, South Africa

Comments: Older SSK (diesel-electric submarine) design, though quite successful. Extensively upgraded, the class continues in service today.

United Kingdom

Of all the nations that operate submarines, none holds closer institutional and engineering ties to the United States than the United Kingdom. Currently the U.K. sub force is undergoing its own downsizing after several decades of steady growth. Part of this is due to financial constraints, though the entire "V" class of SSNs is being retired rather prematurely because of hydrogen embrittlement of valves and other plumbing fixtures in their propulsion plants. As this book is being written, the British are headed for a force of twelve SSNs (Swiftsure and Trafalgar classes), four SSBNs (the Vanguard class), and four SSKs (diesel-electric submarines) (the Upholder class). Even this force may be whittled down, with the Upholders reportedly being considered for sale to export clients. As to future sub construction, the British would like to build a second batch of Trafalgars powered by the British-built PWR-2 power plant, though only the British Parliament and the Minister of Defence are able to determine whether this will happen.

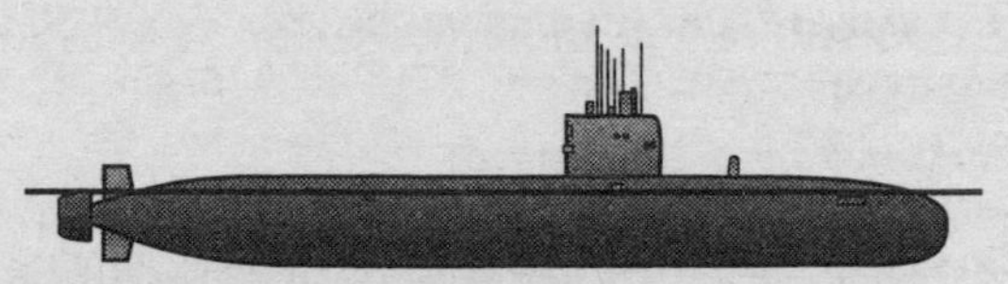

Trafalgar (British). *Jack Ryan Enterprises, Ltd.*

CLASS NAME: Trafalgar

PRODUCER (COUNTRY/MANUFACTURER): United Kingdom/VSEL, Barrow-in-Furness

DISPLACEMENT (SURFACED/SUBMERGED): 4,700/5,208

DIMENSIONS (FT/M): Length: 280.1/85.4 **Beam:** 32.2/9.8 **Draft:** 27.2/8.3

ARMAMENT: Five 533mm torpedo tubes with 25 weapons

MACHINERY: One PWR with steam turbines driving one pumpjet propulsor; 15,000 SHP

SPEED (KNOTS): 30 (submerged)

NUMBER IN CLASS: 7

USERS: United Kingdom

COMMENTS: Quite simply, the best SSN ever built by the British. This class is quick, quiet, and carries a substantial punch. If the Trafalgars have a weakness, it is the lack of an integrated combat system like the U.S. AN/BSY-1. A future upgrade known as Type 2076 may cure this. They handle well and are quite good boats for the money.

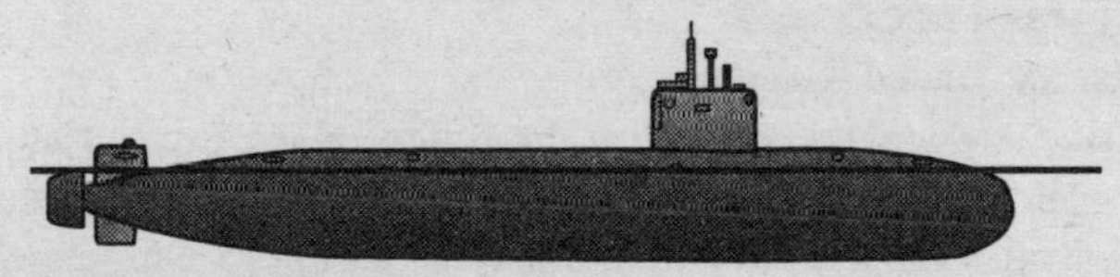

Swiftsure (British). *Jack Ryan Enterprises, Ltd.*

CLASS NAME: Swiftsure

PRODUCER (COUNTRY/MANUFACTURER): United Kingdom/VSEL, Barrow-in-Furness

DISPLACEMENT (SURFACED/SUBMERGED): 4,200/4,500
DIMENSIONS (FT/M): Length: 271.9/82.9 **Beam:** 32.2/9.8 **Draft:** 27.2/8.3
ARMAMENT: Five 533mm torpedo tubes with 25 weapons
MACHINERY: One PWR with steam turbines driving one pumpjet propulsor; 15,000 SHP
SPEED (KNOTS): 30 (submerged)
NUMBER IN CLASS: 5
USERS: United Kingdom
COMMENTS: The oldest SSNs in the Royal Navy, the Swiftsures are fine boats that have been upgraded in refits to almost the same standard as the Trafalgars.

CLASS NAME: Trafalgar Batch II (?)
PRODUCER (COUNTRY/MANUFACTURER): United Kingdom/VSEL, Barrow-in-Furness
DISPLACEMENT (SURFACED/SUBMERGED): ≈ 5,200 (submerged)
DIMENSIONS (FT/M): (unknown)
ARMAMENT: Five 533mm torpedo tubes with 30 weapons
MACHINERY: One PWR with steam turbines driving one pumpjet propulsor; 15,000 SHP
SPEED (KNOTS): ≈ 30 (submerged)
NUMBER IN CLASS: ?
USERS: United Kingdom
COMMENTS: The *big* "If" in the future of the Royal Navy. These boats, if they are ever built, will be powered by the British PWR-2 reactor, and may be equipped with cruise missiles.

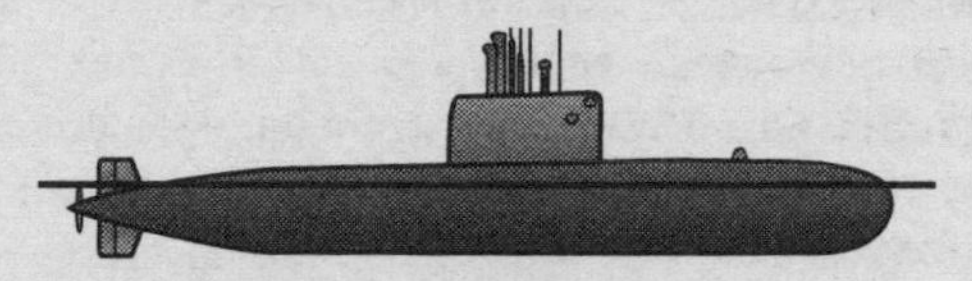

Upholder (British). *Jack Ryan Enterprises, Ltd.*

CLASS NAME: Upholder

PRODUCER (COUNTRY/MANUFACTURER): United Kingdom/VSEL, Barrow-in-Furness; Cammell Laird, Birkenhead

DISPLACEMENT (SURFACED/SUBMERGED): 2,185/2,400

DIMENSIONS (FT/M): Length: 230.6/70.3 **Beam:** 24.9/7.6 **Draft:** 18/5.5

ARMAMENT: Six 533mm torpedo tubes with 18 weapons

MACHINERY: Diesel-electric drive with one 7-bladed screw; 5,400 SHP

SPEED (KNOTS): 20 (submerged)

NUMBER IN CLASS: 4

USERS: United Kingdom

COMMENTS: A *really* nice class of SSKs, these boats are probably the finest diesel-electric submarines in the world. Fully the equal of the Trafalgars in sensors and armament, they may be sold off to export customers.

CLASS NAME: Vanguard

PRODUCER (COUNTRY/MANUFACTURER): United Kingdom/VSEL, Barrow-in-Furness

DISPLACEMENT (SURFACED/SUBMERGED): 15,850 (submerged)

DIMENSIONS (FT/M): Length: 489.7/149.3 **Beam:** 42/12.8 **Draft:** 33.1/10.1

Armament: Sixteen Trident II (D-5) SLBMs; four 533mm torpedo tubes with ≈ 18 weapons

Machinery: One PWR with steam turbines driving one pumpjet propulsor; 27,500 SHP

Speed (knots): 25 (submerged)

Number in class: 1 + 3

Users: United Kingdom

Comments: These are probably going to be the last class of SSBNs ever built for the Royal Navy. Representing everything ever learned by the Royal Navy in submarine design, these elegantly designed boats have something of a "big shoulders" look because of the way the bow planes are mounted.

Resolution (British). *Jack Ryan Enterprises, Ltd.*

Class name: Resolution

Producer (country/manufacturer): United Kingdom/VSEL, Barrow-in-Furness

Displacement (surfaced/submerged): 7,600/8,500

Dimensions (ft/m): Length: 424.8/129.5 **Beam:** 33.1/10.1 **Draft:** 30/9.2

Armament: Sixteen Polaris (A-3) SLBMs; six 533mm torpedo tubes with ≈ 18 weapons

Machinery: One PWR with steam turbines driving one 7-bladed screw; 27,500 SHP

Speed (knots): 25 (submerged)

Number in class: 3

Users: United Kingdom

Comments: The old war-horses of the British submarine force, the units (three remain in commission as of this writing) of the "R" class have been keeping the peace for over a quarter century now. Being retired as the new "V" class boats come on line, they are roughly equivalent to the U.S. Lafayette-class SSBNs.

SWEDEN

Of all the nations that operate submarines, none is probably less understood and more underestimated than Sweden. The Swedes have always had an independent streak when it comes to defense issues, and this is certainly true of their submarine force. At the moment, they produce some of the most advanced conventionally powered submarines in the world. Their boats have a decidedly inshore design philosophy, consistent with the Swedish requirements of operating in the Baltic. In addition, the Swedes are leaders in non-nuclear Air Independent Propulsion (AIP) systems. Currently they are finishing development of the Gotland (A-19 class) boats, equipped with a Sterling AIP system to keep the batteries charged for longer submerged endurance. Like all other nations, the Swedes are aggressively marketing their boats for export. They have had a particular success with the sale of six boats (the Collins class) to Australia.

Gotland (A19)(Swedish). *Jack Ryan Enterprises, Ltd.*

Class name: Gotland (A-19 class)

Producer (country/manufacturer): Sweden/Kockums, Malmö

Displacement (surfaced/submerged): 1,300 (submerged)

Dimensions (ft/m): Length: 172.2/52.5 **Beam:** 19.9/6.1 **Draft:** 18.4/5.6

Armament: Six 533mm and three 400mm torpedo tubes with 18 weapons

Machinery: Diesel-electric drive with one 5-bladed screw; ≈ 4,500 SHP; Sterling engine AIP system to be installed

Speed (knots): 20 (submerged)

Number in class: 0 + 3

Users: Sweden

Comments: Essentially an updated A-17 class, with improved sensors and combat systems and the incorporation of two Sterling engine-driven generators for improved slow-speed submerged endurance.

Västergötland (A-17)(Swedish). *Jack Ryan Enterprises, Ltd.*

Class name: Västergötland (A-17 class)

Producer (country/manufacturer): Sweden/Kockums, Malmö, and Karlskrona Varvet

Displacement (surfaced/submerged): 1,070/1,140

DIMENSIONS (FT/M): Length: 159.1/48.5 **Beam:** 19.9/6.1 **Draft:** 18.4/5.6

ARMAMENT: Six 533mm and three 400mm torpedo tubes with 18 weapons

MACHINERY: Diesel-electric drive with one 5-bladed screw; ≈ 4,000 SHP

SPEED (KNOTS): 20 (submerged)

NUMBER IN CLASS: 4

USERS: Sweden

COMMENTS: Essentially improved Näckens, these boats are quite capable for Baltic operations.

Näcken (A14)(Swedish).*Jack Ryan Enterprises, Ltd.*

CLASS NAME: Näcken (A-14 class)

PRODUCER (COUNTRY/MANUFACTURER): Sweden/Kockums, Malmö, and Karlskrona Varvet

DISPLACEMENT (SURFACED/SUBMERGED): 1,030/1,125

DIMENSIONS (FT/M): Length: 162.4/49.5 **Beam:** 18.7/5.7 **Draft:** 13.4/4.1

ARMAMENT: Six 533mm and two 400mm torpedo tubes with 12 weapons

MACHINERY: Diesel-electric drive with one 5-bladed screw; ≈ 4,000 SHP

SPEED (KNOTS): 20 (submerged)

NUMBER IN CLASS: 3

USERS: Sweden

COMMENTS: The oldest SSKs in the Swedish Navy. Näcken was the trial submarine for the Sterling engine AIP system, which will be incorporated in the Gotland class.

Netherlands

The Dutch enjoy an outstanding submarine tradition, with particular pride in the numerous enemy sinkings to their credit during World War II. In fact, during the early days of 1942 in the Pacific, the tiny Dutch force actually sank more ships than the entire U.S. sub force. Today the Dutch have an excellent fleet of SSKs and are aggressively trying to market them overseas.

Class name: Walrus
Producer (country/manufacturer): Netherlands/Rotterdamse Droogdok Maatschaooij
Displacement (surfaced/submerged): 2,450/2,800
Dimensions (ft/m): Length: 222.2/67.7 **Beam:** 27.6/8.4 **Draft:** 23/7
Armament: Four 533mm torpedo tubes with 20 weapons
Machinery: Diesel-electric drive with one 5-bladed screw; 5,430 SHP
Speed (knots): 21 (submerged)
Number in class: 1 + 3
Users: Netherlands
Comments: A really nice little class of SSK, the Walrus-class boats have a good balance of weapons, sensors, and endurance. The lead boat of the class suffered a fire during building and was delayed in delivery.

Germany

Of all the nations on earth, none has a stronger submarine combat tradition than Germany. Twice in this century the German U-boat fleets have driven England to the brink of starvation and defeat. Today, though, the U-boats of the modern German Navy are a much more modest force, though they probably reflect the missions they would be required to execute better than their counterparts from the two world wars. The new generation of U-boats are tailored to the coastal waters of the Baltic, with endurance and weapons loads to match. The German boats have proven to be a great success, particularly in export sales. In fact the Type 209 has actually outsold the Russian Kilo class in exports, making it the Volkswagen of conventional submarines. Their newest boats, the Type 212s, are capable of being equipped with a liquid oxygen/hydrogen fuel cell AIP system.

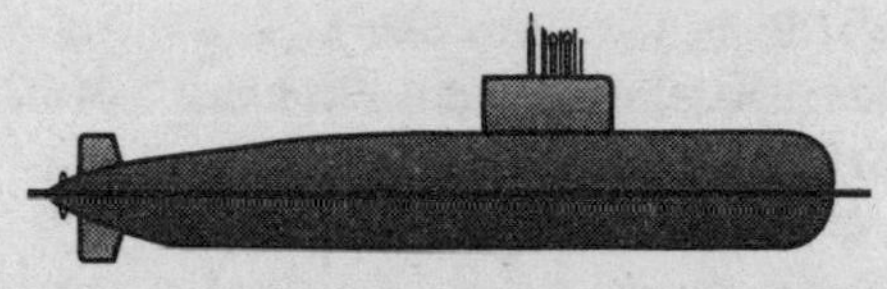

Type 212 (German). *Jack Ryan Enterprises, Ltd.*

CLASS NAME: Type 212
PRODUCER (COUNTRY/MANUFACTURER): Germany/ Howaldtswerke-Deutsche Werft, Thysseen Nordseewerke
DISPLACEMENT (SURFACED/SUBMERGED): 1,200/1,800
DIMENSIONS (FT/M): Length: 167.8/51 **Beam:** 22.6/6.9 **Draft:** 21/6.4
ARMAMENT: Six 533mm torpedo tubes with ≈ 18 weapons
MACHINERY: Diesel-electric drive with one 7-bladed screw; ? SHP; fuel cell AIP system to bc installed
SPEED (KNOTS): ≈ 20 (submerged)
NUMBER IN CLASS: 0 + 12
USERS: Germany
COMMENTS: The newest of the German U-boats. These boats will be equipped with a fuel cell AIP system, though defense cuts could severely affect construction.

Type 206 (German). *Jack Ryan Enterprises, Ltd.*

Class name: Type 206/206A

Producer (country/manufacturer): Germany/ Howaldtswerke-Deutsche Werft, Rheinstahl Nordseewerke

Displacement (surfaced/submerged): 450/520

Dimensions (ft/m): Length: 159.4/48.6 **Beam:** 15.4/4.7 **Draft:** 14.1/4.3

Armament: Eight 533mm torpedo tubes with 16 weapons

Machinery: Diesel-electric drive with one 7-bladed screw; 2,300 SHP

Speed (knots): 17 (submerged)

Number in class: 18

Users: Germany

Comments: Type 206As are modified with the Atlas Electronic CSU 83 integrated sonar suite and the accompanying SLW 83 integrated combat system. Propulsion plant, navigation, and accommodations have also been upgraded.

Type 205 (German). *Jack Ryan Enterprises, Ltd.*

Class name: Type 205

Producer (country/manufacturer): Germany/ Howaldtswerke-Deutsche Werft

Displacement (surfaced/submerged): 419/455

Dimensions (ft/m): Length: 142.7/43.5 **Beam:** 15.1/4.6 **Draft:** 12.5/3.8

Armament: Eight 533mm torpedo tubes with 8 weapons

Machinery: Diesel-electric drive with one 7-bladed screw; 2,300 SHP

Speed (knots): 17 (submerged)

Number in class: 5

Users: Germany, Denmark

Comments: An earlier version of the Type 206, these units will probably be sold or retired in the current round of German defense cuts.

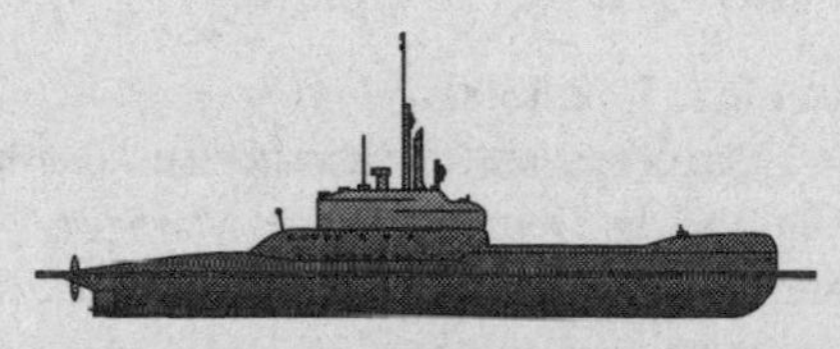

Type 209 (German). *Jack Ryan Enterprises, Ltd.*

Class name: Type 209 (1100, 1200, 1300, 1400 variants)

Producer (country/manufacturer): Germany, Turkey, Brazil, South Korea/various shipyards

Displacement (surfaced/submerged): 1,207–1,586 (submerged)

Dimensions (ft/m): Length: 177.4/54.1–200.7/61.2 **Beam:** 20.5/6.3 **Draft:** 18/5.5

Armament: Eight 533mm torpedo tubes with 14 weapons

Machinery: Diesel-electric drive with one 7-bladed screw; 5,000 SHP

Speed (knots): 22 (submerged)

Number in class: 34 + 15

Users: Argentina, Brazil, Chile, Colombia, Ecuador, Greece, Indonesia, South Korea, Peru, Turkey, Venezuela

Comments: The Type 209 variants differ predominantly in length and displacement, although the sensor, combat, and other electronics fits will also vary depending on when the particular unit was built. Even though the design is over twenty years old, the Type 209 is still being built for customers today, and is the most successful submarine design outside of Russia and the United States.

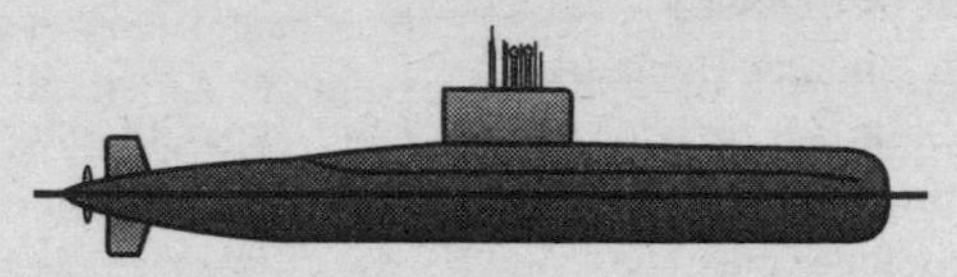

IKL Type 1500 (German). *Jack Ryan Enterprises, Ltd.*

Class name: IKL Type 1500
Producer (country/manufacturer): Germany; India/Howaldtswerke-Deutsche Werft; Mazagon
Displacement (surfaced/submerged): 1,655/1,810
Dimensions (ft/m): Length: 211.2/64.4 **Beam:** 21.3/6.5 **Draft:** 20.3/6.2
Armament: Eight 533mm torpedo tubes with 14 weapons
Machinery: Diesel-electric drive with one 7-bladed screw; 6,100 SHP
Speed (knots): 23 (submerged)
Number in class: 3 + 1
Users: India
Comments: The Type 1500 is normally listed as a Type 209 variant; however, the 1500 has a larger pressure hull and internal compartmentation, which makes it a different design. The internal arrangement of equipment is for all intents and purposes the same as on a Type 209. The Type 1500 is the only Western-designed submarine with an emergency escape sphere in case the boat sinks.

JAPAN

Japan started building its submarine force early; its navy was the first to use subs in combat, during the Russo-Japanese war in the early 1900s. Though Japan produced some of the most advanced boats of World War II, they never really used them to best advantage. Today they operate a large force of SSKs based on the American Barbel class of diesel submarine.

Harushio (Japan). *Jack Ryan Enterprises, Ltd.*

CLASS NAME: Harushio
PRODUCER (COUNTRY/MANUFACTURER): Japan/Mitsubishi
DISPLACEMENT (SURFACED/SUBMERGED): 2,400/2,750
DIMENSIONS (FT/M): Length: 262.4/80 **Beam:** 32.8/10 **Draft:** 25.2/7.7
ARMAMENT: Six 533mm torpedo tubes with 20 weapons
MACHINERY: Diesel-electric drive with one 7-bladed screw; 7,220 SHP
SPEED (KNOTS): 20 (submerged)
NUMBER IN CLASS: 2 + 8
USERS: Japan
COMMENTS: Basically enlarged Yushios, these boats are highly automated, with an excellent weapons load and sensor suite.

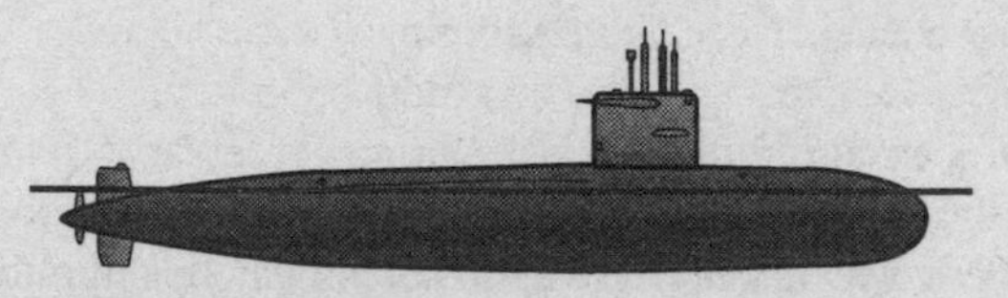

Yushio (Japan). *Jack Ryan Enterprises, Ltd.*

CLASS NAME: Yushio
PRODUCER (COUNTRY/MANUFACTURER): Japan/Mitsubishi and Kawasaki
DISPLACEMENT (SURFACED/SUBMERGED): 2,200/2,450
DIMENSIONS (FT/M): Length: 249.9/76.2 **Beam:** 32.5/9.9 **Draft:** 24.3/7.4
ARMAMENT: Six 533mm torpedo tubes with 20 weapons
MACHINERY: Diesel-electric drive with one 7-bladed screw; 7,220 SHP
SPEED (KNOTS): 20 (submerged)
NUMBER IN CLASS: 10
USERS: Japan
COMMENTS: Very quiet boats armed with both torpedoes and American UGM-84 sub-Harpoon antiship missiles. Capable of very deep operations, the Yushios are updates of the earlier Uzushio class.

ITALY

While the uninformed might not think of Italy as a power in the submarine world, this would be a severe underestimation. Italy has a long and proud history of submarine design, construction, and operations. In World War II, Italian subs did a *lot* of damage to Allied shipping, particularly in the tight waters of the Mediterranean. Following the war, Italy began to build up a substantial force of diesel-electric boats with units constructed in their own yards. Today it is an extremely capable force, which continues to be upgraded with the finest weapons and sensors produced in Italy.

Class name: Primo Longobardo
Producer (country/manufacturer): Italy/Italcantieri
Displacement (surfaced/submerged): 1,653/1,862
Dimensions (ft/m): Length: 217.6/66.4 **Beam:** 22.4/6.8 **Draft:** 19.7/6
Armament: Six 533mm torpedo tubes with 12 weapons
Machinery: Diesel-electric drive with one 7-bladed screw; 4.270 SHP
Speed (knots): 19 (submerged)
Number in class: 0 + 2
Users: Italy
Comments: The Primo Longobardo class is the second modification to the Nazario Sauro class. The biggest differences are in the improved hull form and combat system.

Salvatore Pelosi (Italian). *Jack Ryan Enterprises, Ltd.*

Class name: Salvatore Pelosi
Producer (country/manufacturer): Italy/Fincantiere
Displacement (surfaced/submerged): 1,476/1,662
Dimensions (ft/m): Length: 211.1/64.36 **Beam:** 22.4/6.8 **Draft:** 18.6/5.7
Armament: Six 533mm torpedo tubes with 12 weapons
Machinery: Diesel-electric drive with one 7-bladed screw; 4,270 SHP
Speed (knots): 19 (submerged)
Number in class: 2

USERS: Italy

COMMENTS: Slight hull modifications over the Nazario Sauro class. The Salvatore Pelosi class also has an improved combat system to allow the launch of sub-Harpoon missiles.

Nazario Sauro (Italian). *Jack Ryan Enterprises, Ltd.*

CLASS NAME: Nazario Sauro

PRODUCER (COUNTRY/MANUFACTURER): Italy/C.R.D.A and Italcantiere

DISPLACEMENT (SURFACED/SUBMERGED): 1,450/1,637

DIMENSIONS (FT/M): Length: 209.4/63.9 **Beam:** 22.4/6.8 **Draft:** 18.7/5.7

ARMAMENT: Six 533mm torpedo tubes with 12 weapons

MACHINERY: Diesel-electric drive with one 7-bladed screw; 4,270 SHP

SPEED (KNOTS): 19 (submerged)

NUMBER IN CLASS: 4

USERS: Italy

COMMENTS: Basically improved versions of the Enrico Toti boats that preceded them. Nice little boats designed for operations in the narrows of the seas surrounding Italy.

CLASS NAME: S 90

PRODUCER (COUNTRY/MANUFACTURER): Italy/Fincantiere

DISPLACEMENT (SURFACED/SUBMERGED): 2,500/2,780

DIMENSIONS (FT/M): Length: 228.6/69.7 **Beam:** 26.7/8.2 **Draft:** 20.7/6.3

ARMAMENT: Six 533mm torpedo tubes with 24 weapons

MACHINERY: Diesel-electric drive with one 7-bladed screw; ? SHP

SPEED (KNOTS): 19 (submerged)

NUMBER IN CLASS: 0 + 2

USERS: Italy

COMMENTS: Follow-on to the Primo Longobardo class. Design differences include greater endurance and depth capability. Design not finalized, and changes could occur.

Glossary

1MC Main shipwide announcing circuit on U.S. submarines.

ADCAP ADvanced CAPability. Newest version of the Mark 48 torpedo on board U.S. submarines.

AFFF Aqueous Fire Fighting Foam.

Akula SSN A third-generation Russian design competing with the Sierra I and II classes, the Akula appears to be the overall winner. This boat is very quiet, equivalent to a U.S. Flight I 688, and is equipped with acoustic and nonacoustic sensors. Largest SSN class in production. How large the class size will be is unknown, but at present there are seven Akulas in the Russian inventory.

AN/BPS-15A Navigation radar on many U.S. SSNs.

AN/BQQ-5 (A–E) Integrated sonar suite on most U.S. SSNs. The different variants include improvements in signal process and/or different sonar arrays.

AN/BSY-1 Integrated sonar and fire control system on Improved Los Angeles–class SSNs.

AN/WLR-8(V)2 Radar warning receiver on 688I-class SSNs.

AN/WLR-9 Acoustic intercept receiver found on U.S. Navy submarines.

AN/WLR-10 Radar warning receiver with recording capability on 688-class SSNs.

Anechoic coating Rubber coating applied to the exterior hull surfaces of a submarine to absorb active sonar pulses. Reduces the detectability by active sonars. Some coatings also reduce the amount of noise a submarine puts into the water; these are called decoupling coatings.

Angles and dangles Test conducted by a submarine to ensure that everything is stowed properly before beginning its mission. The procedure calls for making large up-and-down movements with the submarine as well as using large rudder angles at moderate speeds.

ASDIC Allied Submarine Detection Investigation Committee. Formed during World War I (1914–18) to conduct research and experiments on submarine detection.

ASW AntiSubmarine Warfare.

AUTEC Atlantic Undersea Test and Evaluation Center. An acoustic test range located off Andros Island in the Bahamas.

Bastions Highly defended SSBN patrol areas. Established by the former Soviet Union, now used by Russia to protect their SSBNs from attack by Western SSNs.

Blue/gold crew The policy of having two alternating crews aboard strategic missile submarines.

BOL (Bearing Only Launch) Launch mode for Harpoon and Tomahawk antiship missiles that doesn't require range information. Essentially the missile seeker is activated once cruising altitude is reached.

Bomb shop Royal Navy term for the torpedo room on submarines.

BOMBERS Royal Navy nickname for strategic missile submarines.

BOOMERS U.S. Navy nickname for strategic missile submarines.

BOTTOM BOUNCE Term used to describe the route taken by sound waves as they bounce off the ocean bottom traveling from the noise source to the sonar receiver. For example, the noise source could be an active sonar pulse that bounces off the bottom and hits the target ship; then the echo bounces off the bottom again and is received by the sonar.

BREECH DOOR Inner door of a torpedo tube.

BRIDGE Small observation area on top of the fairwater. The OOD stands his watch here when the submarine is on the surface.

BUTTERCUP U.S. Navy term for the "wet" or flooding trainer.

CAVITATION The formation of tiny vapor (air) bubbles on the surface of a propeller when the propeller moves through the water rapidly. Cavitation is a source of very loud noise.

CENTCOM U.S. CENTral COMmand.

CH 084 Multifunction attack periscope on Royal Navy SSNs.

CHOKE POINT Geographical restriction that limits the maneuverability of a ship or submarine.

CIS Commonwealth of Independent States (formerly the Soviet Union).

CK 034 Multifunction search periscope on Royal Navy SSNs.

CLYDE U.S. Navy nickname for the auxiliary diesel engine.

CO Commanding Officer. Title given to an officer in command of a ship. Often called "Captain" or "Skipper."

COB Chief of the Boat. Senior enlisted man in the submarine's crew. Usually a senior or master chief petty officer. Interfaces directly with the XO on issues that affect the enlisted personnel. The Royal Navy equivalent is the coxswain.

COMINT COMmunications INTelligence.

COMSUBLANT COMmander, SUBmarine Force AtLANTic.

COMSUBPAC COMmander, SUBmarine Force PACific.

CONFORM Name of a Navsea-designed SSN in competition with Admiral Rickover's 688 design.

CONTROL ROOM Area on a U.S. Navy submarine where the submarine's ship control, fire control, and periscopes are located. All major submarine functions are controlled from this location. The OOD stands his watch here when the submarine is submerged. In communications the area is referred to as the conn.

CONVERGENCE ZONE (CZ) Phenomenon whereby, if the water is deep enough, water pressure turns sound waves up toward the surface. This occurs at intervals of roughly 30 nautical miles. Multiple CZ contacts are possible when the sound bounces off the surface and heads back down, eventually to be turned back upward again by the pressure.

COW Chief of the Watch. Leading enlisted man in control during a watch. Operates the ballast control panel to dive and surface the submarine and makes trim corrections when directed by the diving officer.

CVBG Aircraft Carrier Battle Group.

DELTA I TO IV SSBNs The Russian Delta series of SSBNs is an ongoing variant of the basic Yankee-class SSBN design. Almost all variations are determined by the type of SLBM being carried. The latest variant, the

Delta IV, also incorporates quieting and sensor enhancements. A total of forty-three Deltas have been built.

Direct path Term used to describe the route that sound waves take from noise source to sonar system without interacting with the surface or the ocean floor. Roughly speaking, it is considered to be the straight-line distance between the two vessels.

DNR Director, Naval Reactors

Dolphins Symbol of the submarine force in just about every nation. Also, the badge or pin that designates a sailor as qualified in submarines.

***Dreadnought* (S-98)** First Royal Navy SSN. Essentially a U.S. Skipjack-class back end mated to a Royal Navy front end.

DSMAC Digital Scene-Matching Area Correlation. A second Tomahawk land attack missile navigation system used to improve the accuracy of the conventional variants. Uses a cameralike system to make detailed digital pictures of the terrain and compares them with stored images in the guidance computer.

DSRV Deep-Submergence Rescue Vehicle. A small rescue submersible designed to dock with a sunken submarine and retrieve the crew.

EAB Emergency Air Breathing system. A low-pressure air system that crewmen can plug in to and obtain breathable—although dry—air. This system is to provide a source of air while a submarine ventilates to get rid of the smoke from a fire.

Echo SSN A first-generation Soviet nuclear-powered submarine, it originally was designed as an SSGN (Echo I class), but the tubes were removed and the units converted to SSNs. These submarines were noisy and had extremely unsafe radiation problems. All have been retired because of their poor safety record. A total of six units were built.

ELECTRIC BOAT COMPANY The company started by John Holland to produce submarines for the U.S. Navy. Presently owned by General Dynamics Corporation.

ELF Extremely Low Frequency radio band.

EMERGENCY BLOW Process by which high-pressure air is rapidly introduced directly into the submarine's main ballast tanks. An emergency blow makes the submarine positively buoyant, and it will rise to the surface quite quickly. This system was instituted as part of the Subsafe program following the loss of the USS *Thresher*.

ENIGMA World War II German communication cipher (encryption) system.

EOOW Engineering Officer of the Watch. Officer in charge of the team that is monitoring and manipulating the submarine's reactor and propulsion system. Key responsibility is to maintain propulsion in a safe manner.

ESM Electronic Support Measures. A passive receiver system designed to detect radar emissions from aircraft and surface ships.

***ETHAN ALLEN* (SSBN 608)** First U.S. Navy SSBN class designed to carry Polaris missiles. Larger than the George Washington class, the Ethan Allen class has more quieting measures to improve stealth. A total of five units were built.

EXOCET Antiship cruise missile made by the French firm Aerospatiale. Slightly smaller than a Harpoon but just as deadly.

FAIRWATER U.S. Navy term for the sail on a submarine. The Royal Navy uses the term *Fin*.

FAMILYGRAMS Short (forty to fifty words) messages that U.S. Navy submariners receive from family members about once a week while on patrol.

FBM Fleet Ballistic Missile submarine.

FIRST LIEUTENANT The Royal Navy equivalent of a U.S. Navy executive officer. Often referred to as "Number One."

"FLAMING DATUM" A ship that has been hit by a torpedo fired from a submarine. It is the place to begin searching for a submarine, because one is known to be in the area.

***GEORGE WASHINGTON* (SSBN-598)** First U.S. Navy SSBN class. Essentially Skipjack-class SSNs with a hull insert containing sixteen missile tubes for Polaris missiles. A total of five units were built.

GERTRUDE Old WW II phrase used to describe any equipment whose function is underwater communications.

***GLENARD P. LIPSCOMB* (SSN-685)** One-of-a-kind experimental U.S. submarine, basically a Sturgeon-class hull with a second-generation turboelectric drive. Fully combat capable.

GOAT LOCKER U.S. Navy term for the chief's quarters on a submarine.

GPS Global Positioning System. A constellation of Navstar satellites that can very accurately determine the submarine's location.

***HALIBUT* (SSN-587)** Originally designed as an SSGN carrying Regulus land attack missiles, she was reconfigured as an SSN when the Polaris program proved to be successful.

HARPOON (UGM-84) U.S. Navy antiship missile, fired from an SSN's torpedo tube.

HE High Explosive.

HEAD U.S. Navy term for a washroom and toilet.

HF High Frequency.

HMS *DOLPHIN* Royal Navy Submarine School.

***HOLLAND* (SS-1)** First U.S. Navy submarine, designed and built by John Holland.

HOT BUNKING Rotation system whereby two men share a single bunk. While one man is on watch the other is sleeping. When it is time for watch rotation the man coming off watch climbs into a bunk that was just recently vacated and is usually still warm.

HOTEL II & III SSBN First-generation Soviet SSBN. These submarines were noisy and extremely unsafe from a radiological standpoint. All have been retired because of their poor safety record and to meet SALT SSBN tube limitations. The Hotel III SSBN was a trial submarine for the SS-N-8 Sawfly SLBM. Approximately nine units were built.

HUNLEY A Confederate Navy vessel that made history by being the first submarine to sink a surface ship in battle (USS *Housatonic*). Unfortunately, the *Hunley* herself also sank in the attack.

HY-80 High-Yield steel, with a yield strength of 80,000 pounds per square inch.

HY-100 High-Yield steel, with a yield strength of 100,000 pounds per square inch.

KILO SS Latest Russian diesel-electric submarine. The Kilo is a medium-range coastal defense submarine that is being offered on the export market. Using state-of-the-art Russian sensors and torpedoes, the Kilo class compares favorably against older Western designs. Russia has twenty Kilos in their naval order of battle, and approximately fourteen have been sold to various countries.

***LAFAYETTE* (SSBN-616)** Third generation of U.S. Navy SSBNs. Larger and quieter than the Ethan Allen class, the Lafayette class carries the Poseidon C-3 missile. However, twelve units of the Lafayette class were backfitted with the Trident I C-4 system during the 1980s. A total of thirty-one units were built.

LF Low Frequency.

LOFAR LOw-Frequency Analyzing and Recording. Term used to describe the process by which narrowband "tonals" are displayed on a modern sonar system.

Los Angeles **(SSN-688)** Admiral Rickover's high-speed submarine design. Most numerous submarine class in the world with a total of sixty-two units to be built. There are three flights with various improvements:

- Flight 1: SSNs 688–718. Basic Los Angeles class.
- Flight 2: SSNs 719–750. VLS, more powerful reactor core.
- Flight 3: SSNs 751–773. AN/BSY-1, bow planes, improved quieting, under-ice capability.

Maneuvering The reactor and propulsion control area located in the engine room. The EOOW stands his watch here.

MEO Marine Engineering Officer. Royal Navy equivalent of the chief engineer; however, an MEO is not eligible for command.

MF Medium Frequency.

MGU Midcourse Guidance Unit. The inertial navigation system used to guide Harpoon and Tomahawk antiship missiles to their targets.

MIDAS MIne Detection and Avoidance Sonar. New mine-hunting sonar on Improved Los Angeles–class SSNs.

Mk 8 (Mark 8) WW II–era straight-running (nonhoming) torpedo used by the Royal Navy up until about the mid-1980s. Two Mk 8s were responsible for the sinking of the Argentinean light cruiser *General Belgrano*.

Mk 48 (Mods 1–4) Designation of the active homing torpedo used by U.S. SSNs. The various modifications have improvements in wire-guidance capability and allow for deeper depths.

Mk 57 U.S. Navy moored influence mine.

Mk 60 Captor EnCAPsulated TORpedo mine. A deep-water moored acoustic influence mine containing a Mark 46 lightweight torpedo as the payload.

Mk 67 SLMM Submarine-Launched Mobile Mine. An obsolete Mk 37 electric torpedo that has been converted into a mobile bottom influence mine.

***Narwhal* (SSN-671)** Basic Sturgeon-class hull with a natural circulation reactor. One-of-a-kind experimental submarine. Fully combat capable.

NATO North Atlantic Treaty Organization

***Nautilus* (SSN-571)** First nuclear-powered submarine in the world. Commissioned September 30, 1954.

Navsea Naval Sea Systems Command.

NIFTI Navy InFrared Thermal Imager.

November SSN First-generation Soviet SSN. Fast, noisy, and extremely unsafe because of radiation. These SSNs have all been retired because of their poor safety record. A total of fourteen were built. One was lost off Cape Finisterre in April 1970.

OBA Oxygen Breathing Apparatus. A portable system that chemically generates oxygen for about 30 minutes. Used by damage control teams to fight fires.

***Ohio* (SSBN-726)** Fourth generation of U.S. Navy SSBNs. Largest submarines in the fleet; each carries twenty-four Trident I C-4 or Trident II D-5 missiles. Extremely quiet submarines. Essentially 688s with twenty-four missile tubes. A total of twenty were to be built, but because of START and the collapse of the USSR only eighteen units will be completed.

OOD Officer Of the Deck. U.S. Navy officer in charge of directing the submarine's movement and ensuring that essential actions are conducted. Primary responsibility is to keep the submarine out of dangerous situations and to keep the captain informed.

OPNAV Office of the Chief of NAVal OPerations.

ORSE Operational Reactor Safeguards Examination.

OSCAR I & II SSGN Third-generation Soviet SSGN, the Oscar is the largest attack submarine ever built. Fast, quiet, and extremely well armed, the Oscar I & II classes are a threat to any surface ship. To date nine units have been built, and production appears to be continuing.

OTTO FUEL The monopropellant (oxidizer and fuel combined) used in Mk 48 and Spearfish torpedoes.

PERISHER Royal Navy Submarine Command Course.

***PERMIT* (SSN-594)** First U.S. Navy production SSN with a primary ASW function. Class was renamed following the loss of USS *Thresher* in April 1963. Fourteen units of this class were eventually built.

PLANK OWNERS The original crew of a boat at the time of its commissioning.

***POLARIS* (A1–A3)** First generation of U.S. Navy submarine-launched ballistic missiles. The different variants each have improvements in range. The Royal Navy uses Polaris A-3 missiles in their Resolution-class SSBNs.

"POLISHING THE CANNONBALL" An attempt to generate a near-perfect fire control solution that may be totally unnecessary. Polishing the cannonball takes too much time, and the submarine may lose its initiative to the intended target

POSEIDON (C-3) Second-generation U.S. Navy submarine-launched ballistic missile.

PSA Post Shakedown Availability. Maintenance period after a new submarine's initial sea trials are completed.

PWR-1 Pressurized Water Reactor-1. The type of reactor found on all current Royal Navy nuclear-powered submarines except the Vanguard-class SSBNs. PWR-1 is essentially the U.S. S5W reactor design, which was sold to the Royal Navy in 1958.

PWR-2 Pressurized Water Reactor-2. An indigenous reactor design for future Royal Navy nuclear-powered submarines. Presently being installed in the new Vanguard-class SSBNs.

RADAR RAdio Detection And Ranging.

RAFT A large metal frame that supports various rotating parts of machinery such as main engines or turbine generators. Through inertial damping it reduces machinery vibrations that could reach the hull. In other words, it's heavy, and the vibrations are absorbed as they try to move the raft.

RAM Radar-Absorbing Material. A coating designed to absorb radar energy and reduce a target's ability to be detected.

RBL-L Range Bearing Launch—Large. A launch mode of Harpoon and Tomahawk antiship missiles that uses both bearing and range information. The "Large" refers to the size of the area where the missile is to conduct its search.

RBL-S Range Bearing Launch—Small. A launch mode of Harpoon and Tomahawk antiship missiles that uses both bearing and range information. The "Small" refers to the size of the area where the missile is to conduct its search.

***RESOLUTION* (S-22)** First Royal Navy SSBN. Very similar to the U.S. Lafayette-class SSBN, the Resolution class carries sixteen U.S. Polaris A-3 missiles armed with British reentry vehicles. A total of four units were built.

RNSH Royal Navy Sub Harpoon.

RORSAT Russian Radar Ocean Reconnaissance SATellite.

S6G The designation of the pressurized water reactor installed in 688-class SSNs.

SAM Surface-to-Air Missile.

SBS Special Boat Service. The Royal Navy equivalent of the U.S. Navy SEALs.

***Scorpion* (SSN-589)** Second U.S. Navy SSN (Skipjack class) to be lost at sea, sometime in May 1968. Most likely cause appears to be an explosion.

SCRAM Safety Control Reactor Axe Man. Term given to the man at the University of Chicago, where the first nuclear core was tested, who was responsible for cutting the rope holding the control rods should something go wrong. The method of inserting control rods has changed considerably, but the term has been retained. With a rapid insertion of control rods the reactor will be made subcritical and will no longer support a sustained nuclear fission reaction.

SEAL SEa-Air-Land. U.S. Navy special forces/commando units.

Seawolf Second U.S. Navy SSN (SSN 575). It is also the class name for the new SSN 21 submarine presently under construction at Electric Boat Company in Groton, Connecticut.

SHF Super High Frequency.

SHP Shaft HorsePower.

Shutter door The outer door of a torpedo tube.

Sierra I & II SSN Third generation of Soviet SSNs. The Sierras are quiet, deep-diving submarines. The pressure hull is made of titanium, which makes Sierra expensive to build. This is reflected in the fact that only four have been built to date. The shipyard that produces Sierras is reportedly going out of the submarine construction business, so four units may be the total class size.

SIGNAL EJECTOR A small (usually 3-inch) torpedo tube-like system for launching flares, noisemakers, and torpedo decoys.

SINS Ship's Inertial Navigation System. A set of gyroscopes that monitor the submarine's position from an established reference point in space.

***SKATE* (SSN-578)** First U.S. Navy production SSN class; four units total.

***SKIPJACK* (SSN-585)** First U.S. Navy SSN class to use the teardrop hull shape. Fastest SSN in the fleet until the Los Angeles class. Total of six units built.

SLBM Submarine-Launched Ballistic Missile.

SLOT Submarine-Launched One-Way Transmitter.

SNAPS Smith Navigation And Plotting System. The navigation and plotting tables used on Royal Navy ships and submarines.

SNAPSHOT Term used to describe the procedure for launching a torpedo in an emergency situation. In a snapshot the submarine crew doesn't have time to conduct TMA but simply shoots a torpedo down the bearing of an incoming weapon or a close contact. Rapid reaction is the basis of the snapshot mode.

SOAC U.S. Navy Submarine Officers Advanced Course.

SOBC U.S. Navy Submarine Officers Basic Course.

SONAR SOund Navigation And Ranging.

SOSUS SOund SUrveillance System. A series of fixed passive sonar arrays used by NATO to provide early warning of deployments into the open ocean of former Soviet submarines.

SOUND ISOLATION MOUNT Springlike mount that absorbs machinery vibration by being stretched and relaxed. The vibration energy needed to move the mount doesn't reach the hull and therefore can't be transmitted into the ocean. These mounts are usually made of metal and rubber, although the Royal Navy prefers a polymer-type spring mount.

SPEARFISH Royal Navy's torpedo equivalent to the Mk 48 ADCAP. Although noisier than the Tigerfish, the Spearfish is faster, with greater endurance and improved homing logic.

SRA Short Range Attack. A mode of firing the Mk 48 torpedo to accommodate a target that is very close to the attacking ship.

SS Diesel-electric attack submarine.

SSBN Strategic ballistic missile submarine, nuclear powered.

SSGN Nuclear-guided (cruise) missile submarine.

SSM Surface-to-Surface Missile. Also used in reference to antiship cruise missiles.

SSN Attack submarine, nuclear powered.

SSK Diesel-electric submarine, hunter-killer.

SS-N-9 SIREN Antiship cruise missile on Russian Charlie II-class SSGNs. Range is about 60 nautical miles.

SS-N-14 SILEX Russian ASW missile that deploys a torpedo or nuclear depth bomb. Its range is about 30 nautical miles.

SS-N-18 STINGRAY Submarine-launched ballistic missile on Russian Delta III SSBNs.

SS-N-19 SHIPWRECK Antiship cruise missile on Russian Oscar-class SSGNs. Range is about 300 nautical miles.

SS-N-20 STURGEON Submarine-launched ballistic missile on Russian Typhoon SSBNs.

SS-N-23 SKIFF Submarine-launched ballistic missile on Russian Delta IV SSBNs.

START STrategic Arms Reduction Treaty.

STEINKE HOOD Combination breathing device and life preserver used during free ascents from a sunken U.S. submarine.

***STURGEON* (SSN-637)** Follow-on to the U.S. Permit class. The Sturgeon class is a little larger and incorporates additional quieting measures. A total of thirty-seven units built.

SUBGRU SUBmarine GRoUp.

SUBROC SUBmarine ROCket. A submarine-launched ballistic rocket with a nuclear depth bomb payload.

SUBRON U.S. SUBmarine SquadRON.

SUBSAFE Procedural and system changes instituted to increase the safety of U.S. submarines following the loss of the USS *Thresher* (SSN-593) in April 1963.

SURTASS SURveillance Towed Array Sensor System (AN/UQQ-2). Essentially a mobile SOSUS array towed by small Ocean Surveillance Ships (T-AGOS).

***SWIFTSURE* (S-104)** Third generation of Royal Navy SSNs. Improved quieting and sensors over the Valiant class. In the redesigned location of the main conformal array the Swiftsures lost a torpedo tube (five instead of six). A total of six units were built.

TASO Torpedo and Anti-Submarine warfare Officer. Royal Navy term for the junior seaman officer in charge of the submarine's torpedo launching system.

TB-16 (A–D) Standard U.S. Navy SSN "fat line" towed array. The various modifications (mods) allow the submarine to search at higher speeds without degrading performance. The array is stored in a sheath that runs along the hull.

TB-23 First U.S. Navy "thin line" array found on SSNs equipped with AN/BSY-1 and AN/BQQ-5E. This array is about four times longer than the TB-16 series and is stored entirely on a reel located in the aft ballast tank area.

TDU Trash Disposal Unit. A tube that ejects weighted trash cylinders from the bottom of a U.S. submarine.

TEA KETTLE Royal Navy term for the reactor on nuclear-powered submarines.

TERCOM TERrain-COntour Matching. A navigation system on Tomahawk land attack missiles. The system uses the Tomahawk's radar altimeter to make terrain profiles at preselected points along the missile's route. These profiles are compared to a radar reference map to determine if flight corrections are needed.

TEZ Total Exclusion Zone.

***THRESHER* (SSN-593)** Lost April 4, 1963, during deep-diving trials following an overhaul period. The loss caused the U.S. Navy to institute the Subsafe program.

TIGERFISH (MK 24, MODS 0–3) A quiet, electric-powered dual-purpose torpedo in service with the Royal Navy.

TMA Target Motion Analysis. The process by which computers or men determine a target's course, speed, and range so that a torpedo or missile can be fired accurately.

TMPS Theater Mission Planning System. U.S. TMPS centers plan Tomahawk land attack mission routes to various targets around the globe using maps and navigation information provided by the Defense Mapping Agency.

TOMAHAWK (UGM-109) Family of cruise missiles that are launched from standard torpedo tubes or special vertical launch tubes on SSNs. The different variants are:

- Tomahawk antiship missile (TASM)
- Tomahawk land attack missile—nuclear (TLAM-N)

- Tomahawk land attack missile—conventional, HE warhead (TLAM-C)
- Tomahawk land attack missile—conventional, bomblets (TLAM-D)

TORPEDO The self-propelled torpedo was invented by Robert Whitehead, an Englishman, in 1866. Since then, the torpedo has undergone significant improvements in speed, range, and depth. Present-day torpedoes are all homing weapons using either active/passive acoustics or wake sensors.

TOWED ARRAY String of passive hydrophones towed at some distance behind the ship. By separating the hydrophones from the ship, the array was no longer limited by platform noise, thereby increasing detection range. The towed array can also be made as long as necessary to detect sounds with long wavelengths (very low frequency).

***TRAFALGAR* (S-107)** Fourth-generation Royal Navy SSN. Basically a slightly larger Swiftsure to accommodate additional quieting measures. Production has just ended, with a total of seven units being built. A modified Trafalgar design, called Trafalgar Batch II, is being worked on with the cancellation of the SSN 20 ("W" class).

TRIDENT I (C-4) Third generation of U.S. Navy submarine-launched ballistic missiles.

TRIDENT II (D-5) Fourth generation of U.S. Navy submarine-launched ballistic missiles.

***TRITON* (SSN-586)** Only U.S. Navy SSN built with two nuclear reactors. Originally designed as a radar picket submarine, *Triton* made a submerged round-the-world cruise in 1960.

TSO Tactical Systems Officer. Royal Navy term for the junior seaman officer in charge of the submarine's fire control system.

***Tullibee* (SSN-597)** First U.S. Navy SSN with torpedo tubes placed amidships to make room for the large 15-foot spherical sonar array. This design is the basis for all later U.S. SSN designs. *Tullibee* was also fitted with a troublesome turboelectric drive, which earned her the reputation of being a hangar queen, and she was often referred to as "Building 597."

Turtle A semisubmersible craft designed and built by David Bushnell during the American Revolutionary War. It was the first submarine to conduct an attack, albeit unsuccessful, against a hostile surface ship (HMS *Eagle*).

Type 18 A multifunction search periscope found on U.S. SSNs.

Type 2 An optics-only attack scope, reminiscent of WW II periscopes, on U.S. submarines.

Type 2019 Acoustic intercept receiver on Royal Navy submarines.

Type 2020 Active/passive conformal array on Royal Navy SSNs.

Type 2027 A computer processor hooked to the Type 2020 to determine range based on multiple arrival paths of a target's noise. Multipath ranging.

Type 2046 Royal Navy towed array for submarines. This clip-on array is attached to one of the stern planes.

Type 2072 A new flank array for Royal Navy SSNs, to replace the older Type 2007.

***Typhoon* SSBN** The size of a small WW II battle cruiser, the Typhoon is the largest submarine ever built. Very quiet, and equipped with modern sensors. A total of six units have been built.

U-BOAT *Unterseeboot*. The German name for submarines.
UAP ESM system on Royal Navy SSNs.
UHF Ultra High Frequency.
ULTRA Allied special intelligence during World War II, obtained by the interception and decryption of German Enigma communications.
***UPHOLDER* (S-40)** Latest diesel-electric submarine in the Royal Navy. Intended to replace the aging Oberon class SS, the Upholders have been experiencing a number of teething pains including problems with their torpedo tubes. With the collapse of the Soviet Union, only four out of a projected class run of twelve units are to be built.
***VALIANT* (S-102)** Second-generation Royal Navy SSN. Based on the *Dreadnought*, the entire submarine was produced in the United Kingdom. A total of five units built.
VANGUARD Second-generation Royal Navy SSBN. Twice as large as the Resolution class, the Vanguard class will carry sixteen U.S. Trident II (D-5) missiles. Described as being very quiet submarines. A total of four units are expected to be built.
VHF Very High Frequency.
VICTOR I & II SSNs Second-generation Soviet SSNs. Larger, quieter, and better equipped than November class. Victor IIs differ from Victor Is in that the Victor IIs have four 650mm torpedo tubes and are about 16 feet longer. A total of twenty-two units were built.
VICTOR III SSN A further modification of a second-generation design. The Victor III class is the first Soviet SSN that came close to Western standards in terms of quieting and sensors. The teardrop-shaped pod on the rudder is the housing for a towed sonar array. The Vic-

tor III is the most numerous class of SSN in the Russian inventory, with twenty-six units built.

VLF Very Low Frequency.

VLS Vertical Launch System. A set of twelve external tubes located in the number two main ballast tank on SSN 719 and on the Los Angeles–class SSN.

VSEL Vickers Shipbuilding Enterprises, Limited. The U.K. equivalent of Electric Boat Company.

WATERFALL DISPLAY Phrase used to describe the appearance that a modern passive sonar display makes while showing bearing versus time information. A contact will look like a bright line on a CRT against a speckled background of other noise sources.

WEO Weapon Engineering Officer. Royal Navy equivalent of the weapons officer; however, the WEO is not eligible for command.

XO Executive Officer. U.S. Navy term for the second in command of a ship.

Bibliography

Magazines

International Defense Review

Jane's Defense Weekly

Jane's Intelligence Review

Maritime Defense

Morskoy Sbornik

Naval Forces—International Forum for Maritime Power

Naval Institute Proceedings

Navy International

The Submarine Review

Books

Anderson, William R., with Clay Blair, Jr. *Nautilus 90 North.* Tab Books, 1989.

Baker, A. D., ed. *Combat Fleets of the World 1993*. Naval Institute Press, 1993.

Barron, John. *Breaking the Ring*. Houghton Mifflin, 1987.

Blake, Bernard, ed. *Jane's Underwater Warfare Systems 1990–91*. Jane's Information Group, 1990.

Breemer, Jan. *Soviet Submarines—Design, Development and Tactics*. Jane's Information Group, 1989.

Burdic, William S. *Underwater Acoustic System Analysis*. Prentice-Hall, 1984.

Bureau of Naval Personnel. *Principles of Naval Engineering*. U.S. Navy, 1970.

Compton-Hall, Richard. *Submarine Warfare: Monsters and Midgets*. Blandford Press, 1985.

———*Sub vs. Sub—The Tactics and Technology of Underwater Warfare*. Orion Books, 1988.

Crane, Jonathan. *Submarine*. British Broadcasting Corp., 1984.

Crouch, Holmes F. *Nuclear Ship Propulsion*. Cornell Maritime Press, 1960.

Daniel, Donald C. *Anti-submarine Warfare and Superpower Strategic Stability*. University of Illinois Press, 1986.

Dönitz, Karl. *Memoirs: Ten Years and Twenty Days*. Naval Institute Press, 1990.

Dolphin Scholarship Foundation. *Thirty Years of Submarine Humor*. Dolphin Scholarship Foundation, 1992.

Earley, Pete. *Family of Spies*. Bantam, 1988.

Frieden, David R., ed. *Principles of Naval Weapon Systems.* Naval Institute Press, 1985.

Friedman, Norman. *Submarine Design and Development.* Naval Institute Press, 1984.

———*U.S. Naval Weapons—Every Gun, Missile, Mine and Torpedo Used by the U.S. Navy from 1883 to the Present Day*. Naval Institute Press, 1987.

———*Desert Victory: The War for Kuwait*. Naval Institute Press, 1991.

———*The Naval Institute Guide to World Naval Weapon Systems 1991/92*. Naval Institute Press, 1991.

Gabler, Ulrich. *Submarine Design*. Bernard & Graefe Verlag, 1986.

Gates, P. J., and N. M. Lynn. *Ships, Submarines and the Sea.* Brassey's, 1990.

Gerken, Louis. *ASW versus Submarine: Technology Battle.* American Scientific Corp., 1986.

Gillmer, Thomas C., and Bruce Johnson. *Introduction to Naval Architecture*. Naval Institute Press, 1982.

Gray, Edwyn. *The Devil's Device*. Naval Institute Press, 1991.

Hassab, Joseph C. *Underwater Signal and Data Processing*. CRC Press, 1989.

Jordan, John. *Soviet Submarines—1945 to the Present*. Arms & Armor Press, 1989.

Kahn, David. *Seizing the Enigma: The Race to Break the German U-Boat Codes, 1939–1943*. Houghton Mifflin, 1991.

Kaufman, Yogi, and Steve Kaufman. *Silent Chase*. Naval Institute Press, 1989.

Kinsler, Lawrence E., Austin R. Frey, Alan B. Coppens, and James V. Sanders. *Fundamentals of Acoustics*, 3d ed. John Wiley & Sons, 1982.

Kramer, A. W. *Nuclear Propulsion for Merchant Ships*. U.S. Government Printing Office, 1962.

Meisner, Arnold. *U.S. Nuclear Submarines*. Concord Publications, 1990.

Miller, David. *Submarines of the World—A Complete Illustrated History 1888 to the Present*. Orion Books, 1991.

Newhouse, John. *War and Peace in the Nuclear Age*. Alfred A. Knopf, 1988.

Peebles, Curtis. *Guardians: Strategic Reconnaissance Satellites*. Presidio Press, 1987.

Polmar, Norman. *The Naval Institute Guide to the Soviet Navy*. 5th ed. Naval Institute Press, 1991.

Polmar, Norman, and Thomas Allen. *Rickover*. Simon and Schuster, 1982.

Polmar, Norman, and Jurrien Noot. *Submarines of the Russian and Soviet Navies 1718–1990*. Naval Institute Press, 1991.

Preston, Anthony. *Submarines: The History and Evolution of Underwater Fighting Vessels*. Octopus Books, 1975.

Richelson, Jeffery T. *America's Secret Eyes in Space: The U.S. Keyhole Spy Satellite Program*. Harper and Row, 1990.

Ross, Donald. *Mechanics of Underwater Noise*. Peninsula Publishing, 1987.

Sakitt, Mark. *Submarine Warfare in the Arctic: Option or Illusion?* Stanford University Press, 1988.

Schwab, Ernest Louis. *Undersea Warriors—Submarines of the World*. Crescent Books, 1991.

Stefanick, Tom. *Strategic Antisubmarine Warfare and Naval Strategy*. Lexington Books, 1987.

Terraine, John. *Business in Great Waters*. Leo Cooper, Ltd., 1989.

Time-Life Books Staff. *Hunters of the Deep*. Time-Life Books, 1992.

Tyler, Patrick. *Running Critical*. Harper and Row, 1986.

Urick, Robert J. *Principles of Underwater Sound*, 3d ed. McGraw Hill, 1983.

U.S. News and World Report Staff. *Triumph Without Victory*. Random House, 1992.

van der Vat, Dan. *The Pacific Campaign: World War II, The U.S./Japanese Naval War 1941–1945*. Simon and Schuster, 1991.

Brochures

"The Closed Cycle Diesel System." Thyssen Nordseewerke GmbH.

"Encapsulated Harpoon." McDonnell Douglas Corp.

"Harpoon." McDonnell Douglas Corp.

"L'Inflexible." Direction Des Constructions Navales.

"Seahake: Torpedo of the Future." STN Systemtechnik Nord GmbH.

"Steam—Its Generation and Use." Babcock & Wilcox, 1978.

"SSN, Rubis Class." Direction Des Constructions Navales.

"SSN, Rubis Class, Améthyste Batch." Direction Des Constructions Navales.

"Tomahawk—A Total Weapon System." McDonnell Douglas Corp.

"TYPE 1400." Howaldtswerke-Deutsche Werft & Ingenieurkontor Lubeck.

"TYPE TR 1000 Mod Ocean-going Submarine." Thyssen Nordseewerke GmbH.

"TYPE TR 1700 Ocean-going Submarine." Thyssen Nordseewerke GmbH.

"VIMOS (Vibration Monitoring System)." Ferranti-Thomson Sonar Systems, UK Ltd.

Pamphlets

Abels, F. (IKL). "German Submarine Development and Design." SNAME/ASE, 1992.

"Diesel-Electric Submarines and Their Equipment." International Defense Review, 1986.

"A Review of the United States Naval Nuclear Propulsion Program." U.S. Department of Defense & Department of Energy, 1990.

"Submarine Roles in the 1990s and Beyond." Assistant Chief of Naval Operations for Undersea Warfare, 1992.

"U.S. Navy Nuclear Submarine Lineup." General Dynamics—Electric Boat Division.

"Vibration and Shock Mount Handbook." Stock Drive Products.

"Welcome Aboard USS *Miami* (SSN-755)." USS *Miami* SSN 755.

"Welcome, Launching of PCU *Santa Fe* (SSN-763)." PCU *Santa Fe*, 1992.

Games

Bond, Larry. "Harpoon." Game Designers Workshop, 1992.

"Computer Harpoon." Three Sixty Software, 1991.

"Red Storm Rising." MPS Technologies, 1988.

"Submarine." Avalon Hill Company, 1977.

"Wolf Pack." Brøderbund Software, 1990.